Algorithms and Combinatorics 8

Study and Research Texts

G. Halász V.T. Sós (Eds.)

Irregularities of Partitions

With 10 Figures

Springer-Verlag
Berlin Heidelberg New York
London Paris Tokyo

Gábor Halász
Eötvös Loránd University, Department of Analysis
Múzeum krt. 6–8, H-1088 Budapest VIII, Hungary

Vera T. Sós
Mathematical Institute of the Hungarian Academy of Sciences
Reáltanoda u. 13–15, H-1053 Budapest V, Hungary

Mathematics Subject Classification (1980): 10-XX, 05-XX, 51-XX

ISBN-13: 978-3-540-50582-2 e-ISBN-13: 978-3-642-61324-1
DOI:10.1007/13: 978-3-642-61324-1

Library of Congress Cataloging-in-Publication Data
Irregularities of partitions. (Algorithms and combinatorics ; 8)
Papers from the meeting held at Fertod, Hungary from July 7th through 10th, 1986.
Bibliography: p.
1. Partitions (Mathematics)–Congresses. I. Halász, Gábor, 1941-. II. Sós, Vera T.
III. Series.
QA165.I77 1989 512'.2 89-5944
ISBN 0-387-50582-2 (U.S.)

Reprint of the original edition 1989

Printing and binding: Druckhaus Beltz, Hemsbach
2141/3140-543210 – Printed on acid-free paper

Preface

The problem of uniform distribution of sequences initiated by Hardy, Littlewood and Weyl in the 1910's has now become an important part of number theory. This is also true, in relation to combinatorics, of what is called Ramsey-theory, a theory of about the same age going back to Schur. Both concern the distribution of sequences of elements in certain collection of subsets. But it was not known until quite recently that the two are closely interweaving bearing fruits for both. At the same time other fields of mathematics, such as ergodic theory, geometry, information theory, algorithm theory etc. have also joined in. (See the survey articles: V.T.Sós: *Irregularities of partitions*, Lecture Notes Series 82, London Math. Soc., Surveys in Combinatorics, 1983, or J. Beck: *Irregularities of distributions and combinatorics*, Lecture Notes Series 103, London Math. Soc., Surveys in Combinatorics, 1985.)

The meeting held at Fertőd, Hungary from the 7th to 11th of July, 1986 was to emphasize this development by bringing together a few people working on different aspects of this circle of problems. Although combinatorics formed the biggest contingent (see papers 2, 3, 6, 7, 13)
some number theoretic and analytic aspects (see papers 4, 10, 11, 14)
generalization of both (5, 8, 9, 12)
as well as irregularities of distribution in the geometric theory of numbers (1), the most important instrument in bringing about the above combination of ideas are also represented.

We would like to express our thanks to every participant of the colloquium for their contributions to its success.

The Editors

Contents

1. Irregularities of Point Distribution Relative to Convex Polygons

J. Beck and W.W.L. Chen

1. Introduction

Let $U^2 = [0,1]^2$ denote the unit square in $\mathbf{R}^2$. Suppose that $\mathcal{P}$ is a distribution of N points in U^2. For any Lebesgue measurable set A in U^2, denote by $Z[\mathcal{P}; A]$ the number of points $\mathcal{P}$ in A. We are interested in the discrepancy function

$$D[\mathcal{P}; A] = Z[\mathcal{P}; A] - N\mu(A), \tag{1}$$

where μ denotes, as usual, the 2-dimensional Lebesgue measure.

We start with the following remarkable theorem of Schmidt on aligned rectangles (i.e. rectangles with sides parallel to the coordinate axes).

Theorem 1. [**Schmidt** [6]]. *There exists a positive absolute constant c_1 such that for every distribution $\mathcal{P}$ of N points in U^2, there is an aligned rectangle A in U^2 such that*

$$|D[\mathcal{P}; A]| > c_1 \log N.$$

An alternative proof of Theorem 1 was given by Halász [3], using a variation of the orthogonal function method introduced by Roth [5]. On the other hand, using the van der Corput sequence, it is not difficult to show that Theorem 1 is best possible (see Halton [4]).

We now consider the more general problem of convex polygons. Let $l \geq 3$, and let θ_i $(1 \leq i \leq l)$ be real numbers satisfying

$$0 \leq \theta_1 < \theta_2 < \ldots < \theta_l < 2\pi. \tag{2}$$

For $1 \leq i \leq l$, write $\mathbf{e}_i = (\cos\theta_i, \sin\theta_i)$. Denote by $\mathrm{POL}(\theta_1, \ldots, \theta_l)$ the family of closed convex l-gons (polygons of l sides) $A = A(\mathbf{a}_1, \ldots, \mathbf{a}_l) \subset U^2$ whose successive vertices $\mathbf{a}_1, \ldots, \mathbf{a}_l$ satisfy

$$\mathbf{a}_i - \mathbf{a}_{i-1} = |\mathbf{a}_i - \mathbf{a}_{i-1}|\mathbf{e}_i$$

for $i = 1, \ldots, l$ (convention $\mathbf{a}_0 = \mathbf{a}_l$). In other words, the sides of A are at angles $\theta_1, \ldots, \theta_l$ to the positive x_1-axis.

In the first half of this paper, we shall prove the following generalization of Theorem 1.

Theorem 2. *Let $l \geq 3$, and let $\underline{\theta} = (\theta_1, \ldots, \theta_l)$ satisfy (2) and such that $\mathrm{POL}(\underline{\theta}) = \mathrm{POL}(\theta_1, \ldots, \theta_l)$ is non-empty. Then there exists a positive constant $c_2(\underline{\theta})$, depending at most on $\underline{\theta}$, such that for every distribution $\mathcal{P}$ of N points in U^2, there is a closed l-gon A in $\mathrm{POL}(\underline{\theta})$ such that*

$$|D[\mathcal{P}; A]| > c_2(\underline{\theta}) \log N.$$

Note that in case $l = 4$ and $\theta_i = \frac{1}{2}(i-1)\pi$ for $1 \leq i \leq 4$, Theorem 2 reduces to Theorem 1. Apart from this special case, it is not known whether Theorem 2 is best possible. We shall, however, illustrate some of the ideas for upper bounds by giving a proof of the following weaker result.

Theorem 3. *Let $l \geq 3$, and let $\underline{\theta} = (\theta_1, \ldots, \theta_l)$ satisfy (2). Then for any $\varepsilon > 0$, there exists a positive constant $c_3(\underline{\theta}, \varepsilon)$, depending at most on $\underline{\theta}$ and ε, such that for every natural number $N \geq 2$, there exists a distribution $\mathcal{P}$ of N points in U^2 such that uniformly*

$$|D[\mathcal{P}; A]| < c_3(\underline{\theta}, \varepsilon)(\log(N d^2(A)))^{5+\varepsilon}, \tag{3}$$

where $d(A)$ denotes the diameter of A and A is an arbitrary l-gon in $\mathrm{POL}(\underline{\theta})$ with $d(A) \geq 2N^{-\frac{1}{2}}$.

We comment here that the inequality (3) can be somewhat improved (see Corollary 4C of [1] or Corollary 20D of [2]).

Theorem 3 can be deduced from the following stronger result (we study a non-compact and renormalized model).

Theorem 4. *Let $l \geq 2$, and let $\underline{\theta} = (\theta_1, \ldots, \theta_l)$ satisfy $0 \leq \theta_1 < \theta_2 < \ldots < \theta_l < \pi$. Let $\mathbf{e}_i = (\cos\theta_i, \sin\theta_i)$ for $1 \leq i \leq l$, and denote by $\mathrm{POL}^\infty(\underline{\theta})$ the family of convex polygons $A \subset \mathbf{R}^2$ such that every side of A is parallel to one of the given directions $\mathbf{e}_i$. Then for every $\varepsilon > 0$, there exists a positive constant $c_4(l, \varepsilon)$, depending at most on l and ε, and an infinite discrete set $\mathcal{Q} = \mathbf{q}_1, \mathbf{q}_2, \mathbf{q}_3, \ldots$ in $\mathbf{R}^2$ such that for every $A \in \mathrm{POL}^\infty(\underline{\theta})$ with $d(A) \geq 2$,*

$$\Big| \sum_{\mathbf{q}_j \in A} 1 - \mu(A) \Big| < c_4(l, \varepsilon)(\log d(A))^{5+\varepsilon},$$

where $d(A)$ denotes the diameter of A.

We shall discuss Theorem 2 in paragraphs 2-4 and Theorem 4 in paragraphs 5-7.

2. A Geometric Argument

Let $l \geq 3$, and let $\underline{\theta} = (\theta_1, \ldots, \theta_l)$ be given. Suppose that $\mathrm{POL}(\underline{\theta})$ is non-empty. Let $A^* = A^*(\mathbf{a}_1, \ldots, \mathbf{a}_l) \in \mathrm{POL}(\underline{\theta})$ be chosen and fixed. Our subsequent argument will depend on this particular choice of A^*. We can modify A^* as follows:

(I) Let $\mathbf{e}_3 = \alpha_1\mathbf{e}_1 + \alpha_2\mathbf{e}_2$, where the scalars α_1 and α_2 are uniquely defined. Then there exists a positive real number α such that the polygon

$$A(\mathbf{a}_1 + \alpha\alpha_1\mathbf{e}_1, \mathbf{a}_2 + \alpha\mathbf{e}_3, \mathbf{a}_3, \ldots, \mathbf{a}_l) \in \mathrm{POL}(\underline{\theta}).$$

In other words, by replacing the vertices $\mathbf{a}_1$ and $\mathbf{a}_2$ of A^* by $\mathbf{a}_1 + \alpha\alpha_1\mathbf{e}_1$ and $\mathbf{a}_2 + \alpha\mathbf{e}_3$ respectively and keeping the other vertices fixed, we obtain another l-gon in $\mathrm{POL}(\underline{\theta})$.

(II) Next, let $\mathbf{e}_2 = \beta_3\mathbf{e}_3 + \beta_4\mathbf{e}_4$, where the scalars β_3 and β_4 are uniquely defined. Then there exists a positive real number β such that the polygon

$$A(\mathbf{a}_1 + \alpha\alpha_1\mathbf{e}_1, \mathbf{a}_2 + \alpha\mathbf{e}_3 - \beta\mathbf{e}_2, \mathbf{a}_3 - \beta\beta_4\mathbf{e}_4, \mathbf{a}_4, \ldots, \mathbf{a}_l) \in \mathrm{POL}(\underline{\theta}).$$

In other words, by replacing the vertices $\mathbf{a}_2 + \alpha\mathbf{e}_3$ and $\mathbf{a}_3$ of the modified polygon in (I), by $\mathbf{a}_2 + \alpha\mathbf{e}_3 - \beta\mathbf{e}_2$ and $\mathbf{a}_3 - \beta\beta_4\mathbf{e}_4$ respectively and keeping the other vertices fixed, we again obtain another l-gon in $\mathrm{POL}(\underline{\theta})$.

Quite simply, we "push two of the edges of A^* inwards". Now let $\mathbf{y} = (y_1, y_2) \in U^2$ be chosen, and let

(4) $A[\mathbf{y}] = A(\mathbf{a}_1 + y_1\alpha\alpha_1\mathbf{e}_1, \mathbf{a}_2 + y_1\alpha\mathbf{e}_3 - y_2\beta\mathbf{e}_2, \mathbf{a}_3 - y_2\beta\beta_4\mathbf{e}_4, \mathbf{a}_4, \ldots, \mathbf{a}_l).$

Clearly $A[\mathbf{y}] \in \mathrm{POL}(\underline{\theta})$ for every $\mathbf{y} \in U^2$.

Let

(5) $$T = \{\mathbf{x} \in U^2 : \mathbf{x} = \mathbf{a}_2 + y_1\alpha\mathbf{e}_3 - y_2\beta\mathbf{e}_2 \text{ where } y \in U^2\}.$$

Then $T \subset U^2$. To prove Theorem 2, it clearly suffices to show that

$$\sup_{\mathbf{y}\in U^2} |D[\mathcal{P}; A[\mathbf{y}]]| \gg \log N,$$

where the implicit constant in the inequality depends at most on the choice of A^*.

We assume from now on that A^* is chosen and fixed.

3. Halász's Method

Following Halász [3], for every distribution $\mathcal{P}$ of N points in U^2, we shall find an auxiliary function $F(\mathbf{y}) = F(\mathcal{P};\mathbf{y})$ which satisfies certain properties. The method of Halász depends on the inequality

$$\int\limits_{U^2} f(\mathbf{y})D(\mathbf{y})d\mathbf{y} \leq \sup_{\mathbf{y}\in U^2} |D(\mathbf{y})| \int\limits_{U^2} |F(\mathbf{y})|d\mathbf{y}, \tag{6}$$

where $D(\mathbf{y}) = D[\mathcal{P};A[\mathbf{y}]]$. We therefore have to find a lower bound for the left-hand side of (6) and an upper bound for the integral on the right-hand side of (6).

Before we can define the function $F(\mathbf{y})$, we need some basic definitions. For the sake of convenience, let $U_0 = [0,1)$. Any $y \in U_0$ can be written in the form

$$y = \sum_{j=0}^{\infty} \delta_j(y)2^{-j-1},$$

where $\delta_j(y) = 0$ or 1 and such that sequence $\delta_j(y)$ does not end with $1,1,\ldots$. For $r = 0,1,2,\ldots$, let

$$R_r(y) = (-1)^{\delta_r(y)}$$

(these are called the Rademacher functions).

Definition. By an r-interval, we mean an interval of the form $[m2^{-r},(m+1)2^{-r})$, where the integer m satisfies $0 \leq m < 2^r$.

Definition. By an (r,s)-box, we mean a set of the form $I_1 \times I_2$, where I_1 is an r-interval and I_2 is an s-interval.

Let the integer n be chosen to satisfy

$$2N \leq 2^n < 4N, \tag{7}$$

where N is the number of points of the distribution $\mathcal{P}$. For any $(r,n-r)$-box B, we consider the parallelogram

$$\widetilde{B} = \{\mathbf{a}_2 + y_1\alpha\mathbf{e}_3 - y_2\beta\mathbf{e}_2 : \mathbf{y} = (y_1,y_2) \in B\}, \tag{8}$$

and write

$$g_r(\mathbf{y}) = \begin{cases} -R_r(y_1)R_{n-r}(y_2) & (B \cap \mathcal{P} = \emptyset), \\ 0 & (B \cap \mathcal{P} \neq \emptyset). \end{cases} \tag{9}$$

The auxiliary function $F(\mathbf{y})$ is defined by

$$F(\mathbf{y}) = \prod_{r=0}^{n}(1 + cg_r(\mathbf{y})) - 1, \tag{10}$$

where $c < \frac{1}{2}$ is a suitably chosen constant.

Lemma 1. *Suppose that* $0 \leq r_1 < r_2 < \ldots < r_k \leq n$. *Then for any* $(r_k, n-r_1)$*-box* B*, exactly one of the following three conditions hold:*
(i) $g_{r_1}(\mathbf{y}) \ldots g_{r_k}(\mathbf{y}) = R_{r_k}(y_1) R_{n-r_1}(y_2)$; or
(ii) $g_{r_1}(\mathbf{y}) \ldots g_{r_k}(\mathbf{y}) = -R_{r_k}(y_1) R_{n-r_1}(y_2)$; or
(iii) $g_{r_1}(\mathbf{y}) \ldots g_{r_k}(\mathbf{y}) = 0$.

Proof. The lemma follows easily on noting that $R_i(y) R_j(y) = \pm R_j(y)$ whenever $i < j$. □

On integrating, we obtain easily

Lemma 2. *Suppose that* $0 \leq r_1 < r_2 < \ldots < r_k \leq n$. *Then*

$$\int_{U^2} g_{r_1}(\mathbf{y}) \ldots g_{r_k}(\mathbf{y}) d\mathbf{y} = 0.$$

We now consider the integral on the right-hand side of (6). It follows from (10) and Lemma 2 that

$$\int_{U^2} |F(\mathbf{y})| d\mathbf{y} \leq 2.$$

It therefore remains to show that

$$\int_{U^2} F(\mathbf{y}) D(\mathbf{y}) d\mathbf{y} \gg \log N. \tag{11}$$

4. Completion of the Proof of Theorem 2

Note, first of all, that

$$F(\mathbf{y}) = cF_1(\mathbf{y}) + \sum_{k=2}^{n+1} c^k F_k(\mathbf{y}), \tag{12}$$

where

$$F_1(\mathbf{y}) = \sum_{r=0}^{n} g_r(\mathbf{y}), \tag{13}$$

and where, for $k = 2, \ldots, n+1$,

$$F_k(\mathbf{y}) = \sum_{0 \leq r_1 < \ldots < r_k \leq n} g_{r_1}(\mathbf{y}) \ldots g_{r_k}(\mathbf{y}). \tag{14}$$

We shall deduce (11) from the following two lemmas.

For simplicity of notation, we let

$$c_5 = \alpha\beta\sin(\theta_3 - \theta_2) = \mu(T), \tag{15}$$

where T is defined by (5).

Lemma 3. *For every $r = 0, \dots, n$, we have that*

$$\int_{U^2} g_r(\mathbf{y}) D(\mathbf{y}) d\mathbf{y} > 2^{-7} c_5.$$

Lemma 4. *For every $k = 2, \dots, n+1$, we have that*

$$\left| \int_{U^2} F_k(\mathbf{y}) D(\mathbf{y}) d\mathbf{y} \right| \le \sum_{r=0}^{n-k+1} \sum_{h=1}^{n-r} \binom{h-1}{k-2} c_5 2^{-h-3}. \tag{16}$$

Before we prove Lemmas 3 and 4, we first deduce (11). It clearly follows from (13) and Lemma 3 that

$$\int_{U^2} F_1(\mathbf{y}) D(\mathbf{y}) d\mathbf{y} > 2^{-7} c_5 n. \tag{17}$$

On the other hand, it follows from Lemma 4 that

(18)

$$\left| \sum_{k=2}^{n+1} c^k \int_{U^2} F_k(\mathbf{y}) D(\mathbf{y}) d\mathbf{y} \right| \le \sum_{k=2}^{n+1} \sum_{r=0}^{n-k+1} \sum_{h=1}^{n-1} c^k \binom{h-1}{k-2} c_5 2^{-h-3} =$$

$$= \sum_{r=0}^{n-1} \sum_{h=1}^{n-r} \sum_{k=2}^{h+1} c^k \binom{h-1}{k-2} c_5 2^{-h-3} \le c^2 c_5 n \sum_{h=1}^{\infty} 2^{-h-3} \sum_{k=0}^{h-1} \binom{h-1}{k} c^k \le$$

$$\le c^2 c_5 n 2^{-4} \sum_{h=0}^{\infty} \left(\frac{1+c}{2} \right)^h \le c^2 c_5 n 2^{-2}.$$

(11) clearly follows from (7), (12), (17) and (18) if $c = 2^{-6}$. It remains to prove lemmas 3 and 4.

The proofs of both lemmas 3 and 4 are based on the following lemma.

Lemma 5. *Suppose that B is an (r,s)-box in U^2 such that $\widetilde{B} \cap \mathcal{P} = \emptyset$, where $\widetilde{B}$ is the parallelogram defined by (8). Then*

$$\int_B R_r(y_1) R_s(y_2) D(\mathbf{y}) d\mathbf{y} = -c_5 N 2^{-2r-2s-4}.$$

Proof. Let

$$B = [m_1 2^{-r}, (m_1+1)2^{-r}) \times [m_2 2^{-s}, (m_2+1)2^{-s}).$$

If we write

$$B' = [m_1 2^{-r}, (m_1+\frac{1}{2})2^{-r}) \times [m_2 2^{-s}, (m_2+\frac{1}{2})2^{-s}),$$

then clearly

$$\int_B R_r(y_1)R_s(y_2)D(\mathbf{y})d\mathbf{y} =$$

$$= \int_{B'} \sum_{\varepsilon_1=0}^{1}\sum_{\varepsilon_2=0}^{1} (-1)^{\varepsilon_1+\varepsilon_2} D[\mathcal{P}; A[(y_1+\varepsilon_1 2^{-r-1}, y_2+\varepsilon_2 2^{-s-1})]]d\mathbf{y}$$

$$= \int_{B'} D[\mathcal{P}; B(\mathbf{y})]d\mathbf{y},$$

where (see (4))

$B(\mathbf{y}) = \{\mathbf{a}_2 + z_1\alpha\mathbf{e}_3 - z_2\beta\mathbf{e}_2 : y_1 \le z_1 < y_1 + 2^{-r-1} \text{ and } y_2 \le z_2 < y_2 + 2^{-s-1}\}$.

Since $B(\mathbf{y}) \subset \widetilde{B}$ for every $\mathbf{y} \in B'$ and $\widetilde{B} \cap \mathcal{P} = \emptyset$, we have that $Z[\mathcal{P}; B(\mathbf{y})] = 0$, so that in view of (1), we have

$$\int_B R_r(y_1)R_s(y_2)D(\mathbf{y})d\mathbf{y} = -N\int_{B'} \mu(B(\mathbf{y}))d\mathbf{y} = -c_5 N 2^{-2r-2s-4},$$

noting that $\mu(B(\mathbf{y})) = 2^{-r-s-2}\mu(T) = c_5 2^{-r-s-2}$. □

Proof of Lemma 3. Let B be an $(r, n-r)$-box in U^2 such that $\widetilde{B} \cap \mathcal{P} = \emptyset$. Then it follows from (9) and Lemma 5 that

$$\int_B g_r(\mathbf{y})D(\mathbf{y})d\mathbf{y} = c_5 N 2^{-2n-4}. \tag{19}$$

There are 2^n $(r, n-r)$-boxes in U^2, so that in view of (7), at least 2^{n-1} of these $(r, n-r)$-boxes contain no points of $\mathcal{P}$. Combining this with (9) and (19), we conclude that

$$\int_{U^2} g_r(\mathbf{y})D(\mathbf{y})d\mathbf{y} \ge c_5 N 2^{-n-5}.$$

The result follows since $2^n < 4N$. □

Proof of Lemma 4. Consider

$$\int_{U^2} g_{r_1}(\mathbf{y})\dots g_{r_k}(\mathbf{y})D(\mathbf{y})d\mathbf{y},$$

where $0 \le r_1 < \ldots < r_k \le n$. Let B be an $(r_k, n - r_1)$-box. Then by Lemmas 1 and 4, we have that

$$\left|\int_B g_{r_1}(\mathbf{y}) \ldots g_{r_k}(\mathbf{y}) D(\mathbf{y}) d\mathbf{y}\right| = \left|\int_B R_{r_k}(y_1) R_{n-r_1}(y_2) D(\mathbf{y}) d\mathbf{y}\right| \le$$
$$\le c_5 N 2^{-2n-2(r_k-r_1)-4}.$$

Since there are exactly $2^{n+(r_k-r_1)}$ $(r_k, n - r_1)$-boxes in U^2, we have that

$$\left|\int_{U^2} g_{r_1}(\mathbf{y}) \ldots g_{r_k}(\mathbf{y}) D(\mathbf{y}) d\mathbf{y}\right| \le c_5 N 2^{-n-(r_k-r_1)-4} = c_5 N 2^{-n-h-4}$$

where $h = r_k - r_1$. (16) follows on noting that once r_1 and h are chosen, there exactly $\binom{h-1}{k-2}$ ways of choosing $(k-2)$ integers in the interval $(r_1, r_1 + h)$. This completes the proof of Lemma 4. □

5. A Combinatorial Lemma

The proof of Theorem 4 is based on a combination of combinatorial and geometric arguments. The combinatorial part is accumulated in the following lemma.

Lemma 6. *Suppose that $X = \{x_1, x_2, \ldots, x_p\}$ is a finite set. For $i = 1, 2, \ldots$, let $\mathcal{Y}^{(i)} = Y_1^{(i)}, Y_2^{(i)}, \ldots$ be a partition of X; in other words,*

$$X = \bigcup_{j \ge 1} Y_j^{(i)}$$

is a union of mutually disjoint sets $Y_j^{(i)}$. For every $k = 1, \ldots, p$, let us associate with x_k a real number $\alpha_k \in [0, 1]$. Then for every $\varepsilon > 0$, there exist a positive constant $c_6(\varepsilon)$, depending only on ε, and integers $a_k = 0$ or 1 $(1 \le k \le p)$ such that

$$\left|\sum_{x_k \in Y_j^{(i)}} (a_k - \alpha_k)\right| < c_6(\varepsilon) i^{1+\varepsilon}$$

for all $Y_j^{(i)}$ satisfying $i \ge 1$ and $j \ge 1$.

Proof. The construction of the integers $a_k = 0$ or 1 $(1 \le k \le p)$ is based on the well-known result in linear algebra that a system of homogeneous linear equations with more variables than equations has a non-trivial solution. We may assume that $0 < \alpha_k < 1 (1 \le k \le p)$. Let $\underline{\alpha} = (\alpha_1, \alpha_2, \ldots, \alpha_p)$. Obviously

$\underline{\alpha} \in [0,1]^p$. Our task is to define a sequence

$$\underline{\alpha}_0, \underline{\alpha}_1 \underline{\alpha}_2, \ldots, \underline{\alpha}_\nu = (\alpha_{1,\nu}, \alpha_{2,\nu}, \ldots, \alpha_{p,\nu}), \ldots$$

of vectors in $[0,1]^p$ that satisfy the following properties: Let

$$X_\nu = \{x_k \in X : a_{k,\nu} \neq 0 \text{ and } \alpha_{k,\nu} \neq 1\}.$$

Then we need

$$X_{\nu+1} \subset X_\nu, \tag{20}$$

$$\alpha_{k,\nu} = 0 \text{ or } 1 \Rightarrow \alpha_{k,\nu} = \alpha_{k,\nu+1}, \tag{21}$$

and

$$\sum_{x_k \in Y_j^{(i)}} \alpha_{k,\nu} = \sum_{x_k \in Y_j^{(i)}} \alpha_{k,\nu+1} \tag{22}$$

for all $Y_j^{(i)}$ with $\#(Y_j^{(i)} \cap X_\nu) \geq c_6(\varepsilon) i^{1+\varepsilon}$. We shall construct the sequence of vectors $\underline{\alpha}_\nu$ by induction. Let $\underline{\alpha}_0 = \underline{\alpha}$. Suppose now that $\underline{\alpha}_\nu$ has been defined and X_ν (see (20)) is non-empty. We let

$$\mathcal{Y}_\nu = \{Y_j^{(i)} : i \geq 1, j \geq 1 \textit{ and } \#(Y_j^{(i)} \cap X_\nu) \geq c_6(\varepsilon) i^{i+\varepsilon}\}$$

and claim that

$$\#Y_\nu < \#X_\nu. \tag{23}$$

This is clear, since $Y_j^{(i)} \cap Y_k^{(i)} = \emptyset$ whenever $j \neq k$ and so we have

$$\#\mathcal{Y}_\nu = \sum_{i \geq 1} \#\{j : \#(Y_j^{(i)} \cap X_\nu) \geq c_6(\varepsilon) i^{1+\varepsilon}\} < \sum_{i=1}^{\infty} \frac{\#X_\nu}{c_6(\varepsilon) i^{1+\varepsilon}} = \#X_\nu$$

with

$$c_6(\varepsilon) = \sum_{i=1}^{\infty} \frac{1}{i^{1+\varepsilon}} < \infty.$$

We now associate a real variable y_k with each x_k $(1 \leq k \leq p)$, and consider the system of linear equations

$$\sum_{x_k \in Y_j^{(i)} \cap X_\nu} y_k = 0$$

for all $Y_j^{(i)} \in \mathcal{Y}_\nu$, and where $y_k = 0$ for all $x_k \in X \setminus X_\nu$. In view of (23), this system has more variables than equations, and so has a non-trivial solution $\mathbf{y} = (y_1, \ldots, y_p)$. Suppose that t_0 is the largest positive value for which the inequalities

$$0 \leq \alpha_{k,\nu} + t_0 y_k \leq 1 (x_k \in x_\nu)$$

hold. Then let

$$\alpha_{k,\nu+1} = \alpha_{k,\nu} + t_0 y_k \quad (1 \leq k \leq p).$$

(20) clearly holds, in view of the maximality of t_0. (21) follows on noting that $\alpha_{k,\nu} = 0$ or 1 implies $x_k \in X \setminus X_\nu$ and so $y_k = 0$. (22) also follows easily. It now follows from (20) that the sequence $\underline{\alpha}_0, \underline{\alpha}_1, \underline{\alpha}_2, \ldots$ will remain constant after a finite number of steps (s steps, say). Then $X_s = \emptyset$ and the vector $\underline{\alpha}_s$ has coordinates 0 and 1 only. We now let $a_k = \alpha_{k,s} (1 \le k \le p)$. Then it follows from (21) and (22) that for all $Y_j^{(i)} \in \mathcal{Y}^{(i)}$,

$$\left| \sum_{x_k \in Y_j^{(i)}} (a_k - \alpha_k) \right| < c_6(\varepsilon) i^{1+\varepsilon}.$$

This completes the proof of Lemma 6. □

6. A Geometric Lemma

Recall that $\underline{\theta} = (\theta_1, \ldots, \theta_l)$ and $\mathbf{e}_i = (\cos\theta_i, \sin\theta_i)$ for $1 \le i \le l$. We now consider the family $\mathrm{POL}^\infty(\underline{\theta}; x_1, x_2)$ of polygons $A \subset \mathbf{R}^2$ such that every side of A is either parallel to $\mathbf{e}_i$ for some $i = 1, \ldots, l$ or parallel to one of the coordinate axes x_1 or x_2.

Our aim in this section is to approximate the characteristic function of an arbitrary polygon in $\mathrm{POL}^\infty(\underline{\theta}; x_1, x_2)$ by those of some special geometric objects. We therefore have to define these special objects first.

Definition. Suppose that $\mathbf{n} = (n_1, n_2) \in \mathbf{Z}^2$. By a special rectangle of order $\mathbf{n}$, we mean a rectangle of the form

(24) $$[m_1 2^{n_1}, (m_1+1)2^{n_1}) \times [m_2 2^{n_2}, (m_2+1)2^{n_2}),$$

where $\mathbf{m} = (m_1, m_2) \in \mathbf{Z}^2$. We denote by $\mathrm{SR}(\mathbf{n})$ the family of all special rectangles of order $\mathbf{n}$.

Definition. Suppose that $1 \le i \le l$. By a triangle of type i, we mean a triangle with sides parallel to x_1, x_2 and $\mathbf{e}_i$.

Let Δ_i be a triangle of type i, where $1 \le i \le l$. Suppose that $t_i^{(1)}$ and $t_i^{(2)}$ denote respectively the lengths of the sides of Δ_i parallel to x_1 and x_2. Let

$$\lambda_i = \frac{t_i^{(1)}}{t_i^{(2)}},$$

and note that the value of λ_i is independent of the choice of the triangle Δ_i.

For any $i = 1, \ldots, l$ and $n \in \mathbf{Z}$, let $\Lambda(i, n)$ denote the rectangular lattice generated by $(2^n \lambda_i^{\frac{1}{2}}, 0)$ and $(0, 2^n \lambda_i^{-\frac{1}{2}})$, i.e. the lattice of points

$$\mathbf{u}(i, n, \mathbf{m}) = (m_1 2^n \lambda_i^{\frac{1}{2}}, m_2 2^n \lambda_i^{-\frac{1}{2}}) \text{ where } \mathbf{m} = (m_1, m_2) \in \mathbf{Z}^2.$$

For convenience of notation, let $\mathbf{E}_1 = (1,0)$ and $\mathbf{E}_2 = (0,1)$.

Definition. Suppose that $1 \leq i \leq l$ and $n \in \mathbf{Z}$. By a special triangle of type i and order n, we mean a triangle with vertices

$$\mathbf{u}(i,n,\mathbf{m}), \mathbf{u}(i,n,\mathbf{m}+\mathbf{E}_1), \mathbf{u}(i,n,\mathbf{m}+\mathbf{E}_2),$$

or a triangle with vertices

$$\mathbf{u}(i,n,\mathbf{m}), \mathbf{u}(i,n,\mathbf{m}-\mathbf{E}_1), \mathbf{u}(i,n,\mathbf{m}-\mathbf{E}_2),$$

where $\mathbf{m} \in \mathbf{Z}^2$. We denote by $\mathrm{ST}(i,n)$ the family of all special triangles of type i and order n.

Definition. Suppose that $1 \leq i \leq l$ and $j = 1$ or 2. By a parallelogram of type (i,j), we mean a parallelogram with sides parallel to $\mathbf{e}_i$ and x_j.

Consider the unit square U^2. For $1 \leq i \leq l$, let $\psi_i^\star$ denote the linear transformation of determinant 1 represented in matrix notation by

$$\psi_i^\star \begin{pmatrix} x_1 \\ x_2 \end{pmatrix} = \begin{pmatrix} \lambda_i^{\frac{1}{2}} & -\lambda_i^{\frac{1}{2}} \\ 0 & \lambda_i^{-\frac{1}{2}} \end{pmatrix} \begin{pmatrix} x_1 \\ x_2 \end{pmatrix}.$$

Let $P_i^\star = \{\psi_i^\star(\mathbf{x}) : \mathbf{x} \in U^2\}$. It is not difficult to see that $P_i^\star$ is the parallelogram with vertices

$$\mathbf{u}(i,0,\underline{0}), \mathbf{u}(i,0,\mathbf{E}_1), \mathbf{u}(i,0,\mathbf{E}_2), \mathbf{u}(i,0,\mathbf{E}_2-\mathbf{E}_1).$$

Definition. Suppose that $1 \leq i \leq l$ and $n \in \mathbf{Z}^2$. By a special parallelogram of type $(i,1)$ and order $\mathbf{n}$, we mean the image under $\psi_i^\star$ of a special rectangle of the form (24), where $\mathbf{m} = (m_1, m_2) \in \mathbf{Z}^2$. We denote by $\mathrm{SP}(i,1,\mathbf{n})$ the family of all special parallelograms of type $(i,1)$ and order $\mathbf{n}$.

Similarly, for $1 \leq i \leq l$, let $\psi_i^{\star\star}$ denote the linear transformation of determinant 1 represented in matrix notation by

$$\psi_i^{\star\star} \begin{pmatrix} x_1 \\ x_2 \end{pmatrix} = \begin{pmatrix} \lambda_i^{\frac{1}{2}} & 0 \\ -\lambda_i^{-\frac{1}{2}} & \lambda_i^{-\frac{1}{2}} \end{pmatrix} \begin{pmatrix} x_1 \\ x_2 \end{pmatrix}.$$

Let $P_i^{\star\star} = \{\psi_i^{\star\star}(\mathbf{x}) : \mathbf{x} \in U^2\}$. $P_i^{\star\star}$ is then the parallelogram with vertices

$$\mathbf{u}(i,0,\underline{0}), \mathbf{u}(i,0,\mathbf{E}_1), \mathbf{u}(i,0,\mathbf{E}_2), \mathbf{u}(i,0,\mathbf{E}_1-\mathbf{E}_2).$$

Definition. Suppose that $1 \leq i \leq l$ and $\mathbf{n} \in \mathbf{Z}^2$. By a special parallelogram of type $(i,2)$ and order $\mathbf{n}$, we mean the image under $\psi_i^{\star\star}$ of s special rectangle of the form (24), where $\mathbf{m} = (m_1, m_2) \in \mathbf{Z}^2$. We denote by $\mathrm{SP}(i,2,\mathbf{n})$ the family of all special parallelograms of type $(i,2)$ and order $\mathbf{n}$.

We shall frequently refer to special rectangles as special parallelogram of type $(0,0)$.

Our work in this section can be summarized in the following lemma.

Lemma 7. *Suppose that $A \in \mathrm{POL}^\infty(\underline{\theta}; x_1, x_2)$ is arbitrary. Then there exist special triangles $T'_1, \ldots, T'_m$ and $T''_1, \ldots, T''_M$ of types $\in \{i, \ldots, l\}$, special parallelogram $P'_1, \ldots, P'_n$ and $P''_1, \ldots, P''_N$ of types $\in \{(0,0)\} \cup \{(i,j) : 1 \le i \le l$ and $j = 1,2\}$ and signs $\varepsilon'_1 = \pm 1, \ldots, \varepsilon'_m = \pm 1, \varepsilon''_1 = \pm 1, \ldots, \varepsilon''_M = \pm 1, \delta'_1 = \pm 1, \ldots, \delta'_n = \pm 1, \delta''_1 = \pm 1, \ldots, \delta''_N = \pm 1$ such that (χ_B stands for the characteristic function of the set B)*

$$\sum_{\nu=1}^{m} \varepsilon'_\nu \chi_{T'_\nu} + \sum_{\beta=1}^{n} \delta'_\beta \chi_{P'_\beta} \le \chi_A \le \sum_{\nu=1}^{M} \varepsilon''_\nu \chi_{T''_\nu} + \sum_{\beta=1}^{N} \delta''_\beta \chi_{P''_\beta} \tag{25}$$

and

$$\sum_{\nu=1}^{M} \varepsilon''_\nu \mu(T''_\nu) + \sum_{\beta=1}^{N} \delta''_\beta \mu(P''_\beta) - \sum_{\nu=1}^{m} \varepsilon'_\nu \mu(T'_\nu) - \sum_{\beta=1}^{n} \delta'_\beta \mu(P'_\beta) \ll l \log(d(A) + 2).$$

Furthermore, these special objects can be chosen in such a way that

$$\max_{\nu,\beta}\{d(T'_\nu), d(P'_\beta), d(T''_\nu), d(P''_\beta)\} \ll d(A)$$

and the numbers m, M, n and N satisfy

$$\max\{m, M\} \ll l \log(d(A) + 2)$$

and

$$\max\{n, N\} \ll l(\log(d(A) + 2))^3.$$

The first step in the proof of Lemma 7 is to reduce the problem to one of investigating rectangles and triangles.

Lemma 8. *Every $A \in \mathrm{POL}^\infty(\underline{\theta}; x_1, x_2)$ is representable in the form*

$$A = (P_1 \cup P_2 \cup P_3 \cup P_4) \setminus \Big(\Big(\bigcup_{\beta=1}^{q_1} R_\beta\Big) \cup \Big(\bigcup_{\nu=1}^{q_2} T_\nu\Big)\Big),$$

where

(i) *$P_1, \ldots, P_4$ are special rectangles of the same order and $d(P_\alpha) < 3d(A)$ for every $\alpha = 1, \ldots, 4$;*
(ii) *for every $\beta = 1, \ldots, q_1$, R_β is an aligned rectangle and $d(R_\beta) < 5d(A)$;*
(iii) *for every $\nu = 1, \ldots, q_2$, T_ν is a triangle of type $\in \{1, \ldots, l\}$ and $d(T_\nu) \le d(A)$;*
(iv) *$q_1 \le 4l + 8$ and $q_2 \le 4l + 6$; and*
(v) *R_β $(\beta = 1, \ldots, q_1)$ and T_ν $(\nu = 1, \ldots, q_2)$ are pairwise disjoint (in the sense that the intersection has measure zero).*

Proof. For $j = 1,2$, denote the projection of A onto the x_j-axis by $A^{(j)}$, and let $L^{(j)}$ denote the length of the interval $A^{(j)}$. Suppose that $n_j \in \mathbf{Z}$ satisfies $2^{n_j-1} < L^{(j)} \leq 2^{n_j}$. Then $A^{(j)}$ is contained in the union of at most two intervals of the type $[m_j 2^{n_j}, (m_j+1)2^{n_j})$, where $m_j \in \mathbf{Z}$. Let $\mathbf{n} = (n_1, n_2)$. Then A is contained in the union of at most four special rectangles of order $\mathbf{n}$. Denote these rectangles by $P_\alpha (\alpha = 1, \ldots, 4)$ with the convention that they may not be distinct, and note that

$$d(P_\alpha) = (2^{2n_1} + 2^{2n_2})^{\frac{1}{2}} < (4d^2 + 4d^2)^{\frac{1}{2}} < 3d,$$

where $d = d(A)$. Suppose now that $P = P_1 \cup \ldots \cup P_4$. For $j = 1, 2$, denote by $P^{(j)}$ the projection of P onto the x_j-axis. Since A is convex, it has at most $(2l+4)$ vertices. It follows that if we draw a straight line parallel to the x_1-axis through each of these vertices, these lines will give a decomposition of A into at most two triangles and at most $(2l+1)$ trapeziums. Let B denote one of these triangles or trapeziums, and for $j = 1, 2$, let $B^{(j)}$ denote the projection of B onto the x_j-axis. Clearly

$$B^{(1)} \times B^{(2)} = B \cup T' \cup T'',$$

where T' and T'' are disjoint triangles of types $\in \{1, \ldots, l\}$ and with diameters not exceeding $d(A)$. Furthermore,

$$P^{(1)} \times B^{(2)} = (B^{(1)} \times B^{(2)}) \cup R' \cup R''$$

for some disjoint aligned rectangles R' and R'' of diameter not exceeding $((4d)^2 + d^2)^{\frac{1}{2}}$. Clearly $A \subset P^{(1)} \times A^{(2)}$, and $(P^{((1)} \times A^{(2)}) \setminus A$ is a (pairwise disjoint) union of at most $(4l+6)$ triangles of types $\in \{1, \ldots, l\}$ and $(4l+6)$ aligned rectangles. Finally, observe that $P \setminus (P^{(1)} \times A^{(2)})$ is a union of at most two disjoint rectangles of diameter not exceeding $((4d)^2 + (2d)^2)^{\frac{1}{2}}$. This completes the proof of Lemma 8. □

Our next step is clearly to investigate these rectangles and triangles obtained from Lemma 8. We first study the rectangles.

Lemma 9. *Suppose that R is an aligned rectangle.*

(i) There exist mutually disjoint special rectangles $R'_1, \ldots, R'_s$, satisfying $s \ll (\log(\mu(R)+2))^2$. such that

$$\bigcup_{\beta=1}^{s} R'_\beta \subset R$$

and

$$\mu[R \setminus [\bigcup_{\beta=1}^{s} R'_\beta]] \leq 1.$$

(ii) There exist mutually disjoint special rectangles $R''_1, \ldots, R''_4$, where $\mu(R'')_\beta < 4\mu(R)$ for $1 \leq \beta \leq 4$, and mutually disjoint special rectan-

gles $R''_5, \dots, R''_t$, satisfying $t \ll (\log(\mu(R)+2))^2$, such that

$$R \subset (R''_1 \cup R''_2 \cup R''_3 \cup R''_4) \setminus [\bigcup_{\beta=5}^{t} R''_\beta]$$

and

$$\mu[[(R''_1 \cup R''_2 \cup R''_3 \cup R''_4) \setminus [\bigcup_{\beta=5}^{t} R''_\beta]] \setminus R] \leq 1.$$

The proof of Lemma 9 is based on the following simple 1-dimensional result. By a special interval, we mean an interval of the type $[m2^n, (m+1)2^n)$, where $m, n \in \mathbf{Z}$. Note that special rectangles are simply the cartesian products of two special intervals.

Lemma 10. *Suppose that (a,b) is an interval in $\mathbf{R}$. Then for every natural number D, there exist special intervals $I_1, \dots, I_D$ such that*

$$\bigcup_{\alpha=1}^{D} I_\alpha \subset [a,b]$$

and

$$\mu_0[[a,b) \setminus [\bigcup_{\alpha=1}^{D} I_\alpha]] \leq (\frac{7}{8})^D (b-a).$$

Here μ_0 denotes the usual measure on $\mathbf{R}$.

Proof. Choose a longest special interval I_1 int $[a,b)$. We then define I_α for $\alpha \geq 2$ inductively such that (i) I_α is a longest special interval contained in $[a,b) \setminus (I_1, \cup \dots \cup I_{\alpha-1})$; (ii) $I_1 \cup \dots \cup I_\alpha$ is an interval; and (iii) if $[a,b) \setminus (I_1 \cup \dots \cup I_{\alpha-1})$ is a union of two disjoint intervals, then I_α belongs to the longer of the two (any one if of equal length). Clearly $\mu_0(I_1) \geq (b-a)/4$. Indeed, if $n \in \mathbf{Z}$ satisfies $2^{n+1} \leq b-a < 2^{n+2}$, then $2^n > (b-a)/4$ and so there exists $m \in \mathbf{Z}$ such that $[m2^n, (m+1)2^n) \subset [a,b)$. A similar argument will give the inequality $\mu_0(I_\alpha) \geq \mu_0([a,b) \setminus (I_1 \cup \dots \cup I_{\alpha-1}))/8$. Lemma 10 follows easily. □

Proof of Lemma 9. Suppose that $R = [a_1, b_1) \times [a_2, b_2)$. For $j = 1, 2$, we now apply Lemma 10 to the interval $[a_j, b_j)$ and obtain special intervals $I_1^{(j)}, \dots, I_{D_j}^{(j)}$, satisfying $D_j \ll \log(\mu(R)+2)$, such that

$$\bigcup_{\alpha_j=1}^{D_j} I_{\alpha_j}^{(j)} \subset [a_j, b_j)$$

and

$$\mu_0([a_j, b_j) \setminus (I_1^{(j)} \cup \ldots \cup I_{D_j}^{(i)})) \leq \frac{b_j - a_j}{2\mu(R)}.$$

The family of special rectangles

$$I_{\alpha_1}^{(1)} \times I_{\alpha_2}^{(2)} \quad (1 \leq \alpha_1 \leq D_1 \text{ and } 1 \leq \alpha_2 \leq D_2)$$

clearly satisfies the requirements of (i). To prove (ii), first note that for $j = 1, 2$, if $n_j \in \mathbf{Z}$ satisfies $2^{n_j - 1} < a_j \leq 2^{n_j}$, then

$$[a_j, b_j) \subset [m_j 2^{n_j}, (m_j + 2)2^{n_j})$$

for some $m_j \in \mathbf{Z}$. Hence

$$R \subset \bigcup_{\varepsilon_1=0}^{1} \bigcup_{\varepsilon_2=0}^{1} ((m_1 + \varepsilon_1)2^{n_1}, (m_1 + \varepsilon_1 + 1)2^{n_1}) \times$$
$$\times ((m_2 + \varepsilon_2)2^{n_2}, (m_2 + \varepsilon_2 + 1)2^{n_2})) = R_1'' \cup R_1'' \cup R_3'' \cup R_4''$$

say. Obviously $\mu(R_\beta'') < 4\mu(R)$ for $1 \leq \beta \leq 4$. Furthermore, the set $(R_1'' \cup R_2'' \cup R_3'' \cup R_4'') \setminus R$ is the disjoint union of at most four aligned rectangles. Applying (i) to each of these completes the proof of Lemma 9. □

Next, we deal with the triangles obtained from Lemma 8. Note that they are of types $\in \{1, \ldots, l\}$.

Definition. Suppose that $1 \leq i \leq l$. By a nice triangle of type i, we mean a triangle which is the intersection of a special triangle T^* of type i and a half-plane with the boundary parallel to one of the sides of T^*.

Suppose that $1 \leq i \leq l$, and that T is a triangle of type i. Let $T_0 \subset T$ be the largest inscribed special triangle of type i. Extending the edges of T_0 to the boundary of T, we see that T is the disjoint (in the sense of measure) union of T_0 and at most three trapeziums and three parallelograms. Each of these trapeziums is clearly the disjoint union of a nice triangle of type i and a parallelogram. Note that all the parallelograms are of types $(0,0), (i,1)$ or $(i,2)$. To summarize,

(26) T is the disjoint union of one special triangle of type i and at most three nice triangles of type i and six parallelograms of types $(0,0)$, $(i,1)$ or $(i,2)$.

It follows that we have to investigate parallelograms of various types as well as nice triangles of type i.

Recall that special parallelograms of type (i, j) and order $\mathbf{n}$ are obtained from special rectangles of order $\mathbf{n}$ by a linear transformation of determinant 1. The following analogue of Lemma 9 is therefore obvious.

Lemma 11. *Suppose that $1 \leq i \leq l$ and $j = 1$ or 2. Suppose furthermore that P is a parallelogram of type (i, j).*

(i) *There exist mutually disjoint special parallelograms $P'_1, \dots, P'_s$ of type (i,j), satisfying $s \ll (\log(\mu(P)+2))^2$, such that*

$$\bigcup_{\beta=1}^{s} P'_\beta \subset P$$

and

$$\mu[P \setminus [\bigcup_{\beta=1}^{s} P'_\beta]] \leq 1.$$

(ii) *There exist mutually disjoint special parallelograms $P''_1, \dots, P''_4$ of type (i,j), where $\mu(P'')_\beta < 4\mu(P)$ for $1 \leq \beta \leq 4$, and mutually disjoint special parallelograms $P''_5, \dots, P''_t$, satisfying $t \ll (\log(\mu(P)+2))^2$, such that*

$$P \subset (P''_1 \cup P''_2 \cup P''_2 \cup P''_4) \setminus [\bigcup_{\beta=5}^{t} P''_\beta]$$

and

$$\mu[[(P''_1 \cup P''_2 \cup P''_3 \cup P''_4) \setminus [\bigcup_{\beta=5}^{t} P''_\beta]] \setminus P] \leq 1.$$

It remains, therefore, to investigate nice triangles.

Lemma 12. *Suppose that $1 \leq i \leq l$. Suppose further that T is a nice triangle of type i.*

(i) *There exist mutually disjoint special triangles $T'_1, \dots, T'_s$ of type i and parallelograms $P'_1, \dots, P'_s$ of types $\in \{(i,1),(i,2),(0,0)\}$, satisfying $s \ll \log(\mu(T)+2)$, such that*

$$[\bigcup_{\nu=1}^{s} T'_\nu] \cup [\bigcup_{\nu=1}^{s} P'_\nu] \subset T$$

and

$$\mu[T \setminus [[\bigcup_{\nu=1}^{s} T'_\nu] \cup [\bigcup_{\nu=1}^{s} P'_\nu]]] \leq 1.$$

(ii) *There exist a special triangle T''_0 of type i, where $d(T''_0) < 2d(T)$, and mutually disjoint special triangles $T''_1, \dots, T''_t$ of type i and parallelograms $P''_1, \dots, P''_q$ of types $\in \{(i,1),(i,2),(0,0)\}$, satisfying $\max\{t,q\} \ll \log(\mu(T)+2)$, such that*

$$T \subset T''_0 \setminus [[\bigcup_{\nu=1}^{t} T''_\nu] \cup [\bigcup_{\nu=1}^{q} P''_\nu]]$$

and

$$\nu[[T''_0 \setminus [[\bigcup_{\nu=1}^{t} T''_\nu] \cup [\bigcup_{\nu=1}^{q} P''_\nu]]]\ T] \leq 1.$$

Proof. (i) will follow if we can prove that for every natural number D, there exist mutually disjoint special triangles $T_1, \ldots, T_D$ of type i and parallelograms $P_1, \ldots, P_D$ of types $\in \{(i,1),(i,2),(0,0)\}$ such that

$$[\bigcup_{\nu=1}^{D} T_\nu] \cup [\bigcup_{\nu=1}^{D} P_\nu] \subset T \tag{27}$$

and

$$\mu[T \setminus [[\bigcup_{\nu=1}^{D} T_\nu] \cup [\bigcup_{\nu=1}^{D} P_\nu]]] \leq 4^{-D}\mu(T). \tag{28}$$

To prove (27) and (28), note that T, being a nice triangle of type i, is the intersection of a special T^* of type i and a half-plane H with boundary parallel to one of the sides of T. Let $\mathbf{v}'$ and $\mathbf{v}''$ denote the vertices of T on the boundary of H, and let $\mathbf{c}$ denote the third vertex of T. Suppose that $T_1 \subset T$ is the largest inscribed special triangle of type i. Then $\mathbf{c}$ is a vertex of T_1 and $\mu(T_1) \geq \mu(T)/4$. Let $\mathbf{v}_1'$ and $\mathbf{v}_1''$ denote the other two vertices of T_1. The trapezium with vertices $\mathbf{v}', \mathbf{v}'', \mathbf{v}_1', \mathbf{v}_1''$ is then clearly the disjoint union of a nice triangle T_1' and a parallelogram P_1 of type $\in \{(i,1),(i,2),(0,0)\}$. Obviously $\mu(T_1') \leq \mu(T)/4$. We now repeat the argument to T_1' and obtain a special triangle T_2', parallelogram P_2 and nice triangle T_2', mutually disjoint and such that $T_1' = T_2 \cup P_2 \cup T_2'$ and $\mu(T_2') \leq \mu(T_1')/4$. After D steps, we obtain (27) and (28). (i) now follows from a suitable choice of D. To prove (ii), denote by T_0'' the smallest special triangle of type i such that T is the intersection of T_0'' and some half-plane H. Then $d(T_0'') < 2d(T)$. Furthermore, $T_0'' \setminus T$ is the disjoint union of a nice triangle of type i and a parallelogram of type $\in \{(i,1),(i,2),(0,0)\}$. (ii) now follows on applying (i) to this latter nice triangle. This completes the proof of Lemma 12.

□

Lemma 7 follows on combining the observation (26) and Lemmas 8, 9, 11 and 12.

7. Completion of the Proof of Theorem 4

Given any discrete subset $\mathcal{P} \subset \mathbf{R}^2$ and any compact subset $B \subset \mathbf{R}^2$, let $Z[\mathcal{P}; B]$ denote the number of points of $\mathcal{P}$ in B. We are interested in the discrepancy function

$$\mathcal{D}[\mathcal{P}; B] = Z[\mathcal{P}; B] - \mu(B).$$

Suppose that $A \in \mathrm{POL}^{\infty}(\underline{\theta}; x_1, x_2)$ is arbitrary. We first try to use the geometric information derived from Lemma 7 on A to investigate the discrepancy function of A.

The following lemma is in a more general form than needed.

Lemma 13. *Suppose that $A, B'_1, \ldots, B'_q, B''_1, \ldots, B''_r$ are compact subsets of $\mathbf{R}^2$. Suppose further that there exist $\varepsilon'_\alpha = \pm 1$ $(\alpha = 1, \ldots, q)$ and $\varepsilon''_\tau = \pm 1$ $(\tau = 1, \ldots, r)$ such that*

$$\sum_{\alpha=1}^{q} \varepsilon'_\alpha \chi_{B'_\alpha} \leq \chi_A \leq \sum_{\tau=1^r} \varepsilon''_\tau \chi_{B''_\tau}$$

and

$$\sum_{\tau=1}^{r} \varepsilon''_\tau \mu(B''_\tau) - \sum_{\alpha=1}^{q} \varepsilon'_\alpha \mu(B'_\alpha) \leq D_1.$$

Let $\mathcal{P} \subset \mathbf{R}^2$ be a discrete set such that

$$\max_{\alpha,\tau} |\mathcal{D}[\mathcal{P}; B']|, |\mathcal{D}[\mathcal{P}; B'']| \leq D_2.$$

Then

$$|\mathcal{D}[\mathcal{P}; A]| \leq D_1 + D_2 \max\{q, r\}.$$

Proof. Clearly

(29)
$$\begin{aligned} \mathcal{D}[\mathcal{P}; A] &= \sum_{\mathrm{p} \in A \cap \mathcal{P}} 1 - \mu(A) \leq \sum_{\tau=1}^{r} \varepsilon''_\tau \sum_{\mathrm{p} \in B''_\tau \cap \mathcal{P}} 1 - \mu(A) = \\ &= \sum_{\tau=1}^{r} \varepsilon''_\tau \left[\sum_{\mathrm{p} \in B''_\tau \cap \mathcal{P}} 1 - \mu(B''_\tau) \right] + \left[\sum_{\tau=1}^{r} \varepsilon''_\tau \mu(B''_\tau) - \mu(A) \right] = \\ &= \sum_{\tau=1}^{r} \varepsilon''_\tau \mathcal{D}[\mathcal{P}; B''_\tau] + \left[\sum_{\tau=1}^{r} \varepsilon''_\tau \mu(B''_\tau) - \mu(A) \right] \leq \\ &\leq \sum_{\tau=1}^{r} |\mathcal{D}[\mathcal{P}; B''_\tau]| + \left[\sum_{\tau=1}^{r} \varepsilon''_\tau \mu(B''_\tau) - \sum_{\alpha=1}^{q} \varepsilon'_\alpha \mu(B'_\alpha) \right] \leq \\ &\leq D_2 r + D_1. \end{aligned}$$

A similar argument gives

(30)
$$-\mathcal{D}[\mathcal{P}; A] \leq D_2 q + D_1.$$

Lemma 13 follows on combining (29) and (30). □

Let $\mathrm{SPEC}^\infty(\underline{\theta}; x_1, x_2)$ denote the big family of all special triangles, special parallelograms and special rectangles defined in §6, i.e.

$$\mathrm{SPEC}^\infty(\underline{\theta}; x_1, x_2) = \Big[\bigcup_{\substack{1\le i\le l\\ n\in\mathbf{Z}}} ST(i,n)\Big] \cup \Big[\bigcup_{\substack{1\le i\le l\\ 1\le j\le 2\\ \mathbf{n}\in\mathbf{Z}^2}} SP(i,j,\mathbf{n})\Big] \cup \Big[\bigcup_{\mathbf{n}\in\mathbf{Z}^2} \mathrm{SR}(\mathbf{n})\Big].$$

Next, we make use of the combinatorial information derived from Lemma 6.

Lemma 14. *Suppose that $\mathcal{P} \subset \mathbf{R}^2$ is a finite set, and $\alpha \in [0,1]$ is fixed. Then there exists a function $f; \mathcal{P} \to \{-\alpha, 1-\alpha\}$ such that for every polygon $B \in \mathrm{SPEC}^\infty(\underline{\theta}; x_1, x_2)$ satisfying $d(B) \ge 1$, we have*

$$\Big|\sum_{\mathbf{p}\in B\cap\mathcal{P}} f(\mathbf{p})\Big| \ll_\varepsilon (l(\log(d(B)+2))^2)^{1+\varepsilon}.$$

Proof. We apply Lemma 6 with $X = \mathcal{P}$, and so have to introduce a sequence $\mathcal{Y}^{(1)}, \mathcal{Y}^{(2)}, \mathcal{Y}^{(3)}, \ldots$ of partitions of $\mathcal{P}$. Let

$$\begin{aligned}\mathrm{SET}^\infty(\underline{\theta}; x_1, x_2) = &\{\mathrm{ST}(i,n) : 1 \le i \le l \text{ and } n \in \mathbf{Z}\} \cup \\ &\cup \{\mathrm{SP}(i,j,\mathbf{n}) : 1 \le i \le l, 1 \le j \le 2 \text{ and } \mathbf{n} \in \mathbf{Z}^2\} \cup \\ &\cup \{\mathrm{SR}(\mathbf{n}) : \mathbf{n} \in \mathbf{Z}^2\}.\end{aligned}$$

For every $C \in \mathrm{SET}^\infty(\underline{\theta}; x_1, x_2)$, denote by $d(C)$ the common diameter of all the elements of C. We next define a linear ordering on the subset

$$\{C \in \mathrm{SET}^\infty(\underline{\theta}; x_1, x_2) : d(C) \ge 1\}$$

according to the size of $d(C)$ with the convention that this ordering is defined arbitrarily in the case of equal diameters. Observe that for any $y \ge 1$,

$$\begin{aligned}&\#\{C \in \mathrm{SET}^\infty(\underline{\theta}; x_1, x_2) : 1 \le d(C) \le y\} = \\ &= \sum_{i=1}^{l} \#\{n \in \mathbf{Z} : 1 \le d(\mathrm{ST}(i,n)) \le y\} + \\ &+ \sum_{i=1}^{l}\sum_{j=1}^{2} \#\{(\mathbf{n} \in \mathbf{Z}^2 : 1 \le d(\mathrm{SP}(i,j,\mathbf{n})) \le y\} + \\ &+ \#\{\mathbf{n} \in \mathbf{Z}^2 : 1 \le d(\mathrm{SR}(\mathbf{n})) \le y\} \ll \\ &\ll l\log(y+2) + l(\log(y+2))^2 \ll l(\log(y+2))^2.\end{aligned} \tag{31}$$

Suppose that $\mathcal{P}$ is fixed. We now let $\mathcal{Y}^{(1)}, \mathcal{Y}^{(2)}, \mathcal{Y}^{(3)}, \ldots$ be the partitions of $\mathcal{P}$ defined by the families in $\{C \in \mathrm{SET}^\infty(\underline{\theta}; x_1, x_2) : 1 \le d(C) \le d(B)\}$ ordered in the way described. Lemma 14 now follows from Lemma 6 and (31). □

Let $\kappa = 2^k$, where $k \geq 1$ is an integer, and consider the set

$$\mathcal{P} = \{(a/\kappa, b/\kappa) : a, b \in \mathbf{Z} \textit{ and } -\kappa^2 \leq a, b < \kappa^2\}.$$

in the square $[-\kappa, \kappa)^2$. Clearly $\#\mathcal{P} = 4\kappa^4$. Let $\alpha = \kappa^{-2}$. Then $\alpha\#\mathcal{P} = 4\kappa^2$, i.e. the expected number of points of the desired set $\mathcal{Q}$ in $[-\kappa, \kappa)^2$. We now apply Lemma 14. There exists a function $f : \mathcal{P} \to \{-\alpha, 1-\alpha\}$ such that for all polygons $B \in \mathrm{SPEC}^\infty(\underline{\theta}; x_1, x_2)$ satisfying $B \subset [-\kappa, \kappa)^2$ and $d(B) \geq 1$, we have

$$\left|\sum_{\mathbf{p} \in B \cap \mathcal{P}} f(\mathbf{p})\right| \ll_\varepsilon (l(\log(d(B) + 2))^2)^{1+\varepsilon}. \tag{32}$$

Writing $\mathcal{P}_k = \{\mathbf{p} \in \mathcal{P} : f(\mathbf{p}) = 1 - \alpha\}$, we have

$$\sum_{\mathbf{p} \in B \cap \mathcal{P}} f(\mathbf{p}) = \sum_{\mathbf{p} \in B \cap \mathcal{P}_k} 1 - \kappa^{-2} \sum_{\mathbf{p} \in B \cap \mathcal{P}} 1. \tag{33}$$

Furthermore, it is easy to see that for any convex $B \subset [-\kappa, \kappa)^2$

$$\left|\sum_{\mathbf{p} \in B \cap \mathcal{P}} 1 - \kappa^2 \mu(B)\right| \ll \kappa\sigma(\partial B) \ll \kappa^2, \tag{34}$$

where $\sigma(\partial B)$ denotes the length of the perimeter of B. It follows, on combining (32)–(34), that

$$|\mathcal{D}[\mathcal{P}_k; B]| =$$

$$= \left|\sum_{\mathbf{p} \in B \cap \mathcal{P}_k} 1 - \mu(B)\right| \leq \left|\sum_{\mathbf{p} \in B \cap \mathcal{P}_k} 1 - \kappa^{-2} \sum_{\mathbf{p} \in B \cap \mathcal{P}} 1\right| + \left|\kappa^{-2} \sum_{\mathbf{p} \in B \cap \mathcal{P}} 1 - \mu(B)\right| \ll$$

$$\ll_\varepsilon (l(\log(d(B) + 2))^2)^{1+\varepsilon}$$

for all $B \in \mathrm{SPEC}^\infty(\underline{\theta}; x_1, x_2)$ satisfying $B \subset [-\kappa, \kappa)^2$ and $d(B) \geq 1$.

Suppose now that $B \in \mathrm{SPEC}^\infty(\underline{\theta}; x_1, x_2)$ satisfies $B \subset [-\kappa, \kappa)^2$ and $d(B) < 1$. Then $B \subset B_0$ for some $B_0 \in \mathrm{SPEC}^\infty(\underline{\theta}; x_1, x_2)$ with $1 \leq d(B_0) < 2$. Applying (32) and (33) to B_0, we have

$$\sum_{\mathbf{p} \in B_0 \cap \mathcal{P}_k} 1 = \kappa^{-2} \sum_{\mathbf{p} \in B_0 \cap \mathcal{P}} 1 + \sum_{\mathbf{p} \in B_0 \cap \mathcal{P}} f(\mathbf{p}) = 4 + \sum_{\mathbf{p} \in B_0 \cap \mathcal{P}} f(\mathbf{p}) \ll_\varepsilon l^{1+\varepsilon},$$

noting that $\mu(B_0) \leq (d(B_0))^2 < 4$. Hence

$$\sum_{\mathbf{p} \in B \cap \mathcal{P}_k} 1 \leq \sum_{\mathbf{p} \in B_0 \cap \mathcal{P}_k} 1 \ll_\varepsilon (l(\log(d(B) + 2))^2)^{1+\varepsilon}.$$

Using $\mu(B) < \mu(B_0) < 4$, we have

$$|\mathcal{D}[\mathcal{P}_k; B]| \ll_\varepsilon (l(\log(d(B) + 2))^2)^{1+\varepsilon}. \tag{35}$$

It now follows that (35) holds for all $B \in \mathrm{SPEC}^\infty(\underline{\theta}; x_1, x_2)$ satisfying $B \subset [-\kappa, \kappa)^2$. Combining this with Lemmas 7 and 13, we conclude that

$$|\mathcal{D}[\mathcal{P}_k; C]| \ll l^{2+\varepsilon}(\log(d(C) + 2))^{5+\varepsilon} \tag{36}$$

for all $C \in \mathrm{POL}^\infty(\underline{\theta}; x_1, x_2)$ satisfying $C \subset [-\kappa, \kappa)^2$.

At last, we are in a position to construct the set $\mathcal{Q}$ in terms of the sets $\mathcal{P}_k$. Note first of all that

$$\bigcup_{n\in\mathbf{N}} [[-2^{2^n}, 2^{2^n}]^2 \setminus [-2^{2^{n-1}}, 2^{2^{n-1}}]^2] = \mathbf{R}^2 \setminus [-2, 2)^2,$$

and that any set in this union is the disjoint union of four aligned rectangles. We shall show that the set

$$\mathcal{Q} = \mathcal{P}_1 \cup [\bigcup_{\substack{k=2^n\\ n\in\mathbf{N}}} (\mathcal{P}_k \cap ([-2^k, 2^k)^2 \setminus [-2^{\frac{1}{2}k}, 2^{\frac{1}{2}k})^2))]$$

satisfies the requirements of Theorem 4. Consider any arbitrary $A \in \mathrm{POL}^\infty(\underline{\theta}; x_1, x_2)$. The intersection

$$A_k = A \cap ([-2^k, 2^k)^2 \setminus [-2^{\frac{1}{2}k}, 2^{\frac{1}{2}k})^2)$$

is the disjoint union of at most four sets in $\mathrm{POL}^\infty(\underline{\theta}; x_1, x_2)$, so that by (36), we have

(37)

$$|D[\mathcal{Q}; A]| = \left| \sum_{q \in a \cap \mathcal{Q}} 1 - \mu(A) \right| =$$

$$= \left| \#(A \cap \mathcal{P}_1) - \mu(A \cap [-2, 2)^2 + \sum_{\substack{k=2^n\\ n\in\mathbf{N}}} (\#(A_k \cap \mathcal{P}_k) - \mu(A_k)) \right| \ll_\varepsilon$$

$$\ll_\varepsilon \sum\nolimits^* l^{2+\varepsilon} (\min\{\log(d(A) + 2), k\})^{5+\varepsilon}.$$

Here the summation $\sum^*$ is extended over all $k = 2^n$, where $n \in \mathbf{N}$, for which A_k is non-empty. Simple calculation gives

$$\sum\nolimits^* (\min\{\log(d(A) + 2), k\})^{5+\varepsilon} \ll (\log(d(A) + 2)^{5+\varepsilon}. \tag{38}$$

Theorem 4 now follows from (37) and (38). □

References

[1] J. Beck, Irregularities of distribution II (to appear in the London Mathematical Society).

[2] J. Beck and W.W.L. Chen, *Irregularities of distribution* (Cambridge Tracts in Mathematics **89**, Cambridge University Press, 1987).

[3] H. Halász, On Roth's method in the theory of irregularities of point distributions, *Recent progress in analytic number theory*, **2**, 79-94 (Academic Press, 1981).

[4] J.H. Halton, On the efficiency of certain quasirandom sequences of points in evaluating multidimensional integrals, *Num. Math.*, **2** (1960), 84-90.

[5] K.F. Roth, On irregularities of distribution, *Mathematika*, **1** (1954), 73-79.

[6] W.M.Schmidt, Irregularities of distribution VII, *Acta Arith.*, **21** (1972), 45-50.

J. Beck
L. Eötvös University
Budapest

W.W.L. Chen
Imperial College
London

2. Balancing Matrices with Line Shifts II

J. Beck and J. Spencer

1. Statement and Reductions

Let an arbitrary matrix $A = (a_{ij})$, $1 \le i \le K$, $1 \le j \le L$ be given with all $|a_{ij}| \le 1$. By a row shift we mean the act of replacing, for a particular i, all coefficients a_{ij} in the i-th row by their negatives $(-a_{ij})$. A column shift is defined similarly. A line shift denotes either a row or a column shift. Consider the following solitaire game. The player applies a succession of line shifts to A. His object is to make the absolute value of the sum of all the coefficients of A (which we shall denote by $|A|$) as small as possible. We shall show (answering a question of J. Komlós) that the player can always make $|A| \le c_0$ where c_0 is an absolute constant — i.e., independent of K, L, and the initial matrix. We make no attempt to find the minimal possible c_0.

Where $a_{ij} \in \{-1, +1\}$ for all i, j this game has been considered by several authors. J. Komlós and M. Sulyok [3], resolving a conjecture of Leo Moser, showed that if $K = L$ and is sufficiently large then $|A| \le 2$ may be achieved. ($|A| \le 1$ if K is odd.) We gave a different proof of this result (for all K) in our earlier paper [1]. We make here extensive use of methods used in these previous works.

We observe that line shifts are commutative and of order two. Thus we may consider a set, rather then a sequence, of line shifts. We restate our results more formally.

Theorem 1. *There is a constant c_0 with the following property. Let $A = (a_{ij})$ be a $K \times L$ matrix (K, L arbitrary) with all $|a_{ij}| \le 1$. Then there exist $\varepsilon_1, \ldots, \varepsilon_K$, $\delta_1, \ldots, \delta_L \in \{-1, +1\}$ so that*

$$(1) \qquad \left| \sum_{i=1}^{K} \sum_{j=1}^{L} \varepsilon_i \delta_j a_{ij} \right| \le c_0.$$

Our plan, to oversimplify somewhat, is to find a set of column shifts so that the row sums have certain properties. Row shifts will then be found by the following result.

Lemma 2. *Let $r_1, \dots, r_K$ be such that*

$$c_0 + |r_1| + \dots + |r_{i-1}| \geq |r_i| \text{ for } 1 \leq i \leq K. \tag{2}$$

(For $i = 1$ this means $c_0 \geq |r_1|$.) Then there exist $\varepsilon_1, \dots, \varepsilon_K \in \{-1, +1\}$ so that

$$|\varepsilon_1 r_1 + \varepsilon_2 r_2 + \dots + \varepsilon_K r_K| \leq c_0. \tag{3}$$

Proof (Outline). We set $\varepsilon_K = +1$ and define ε_i in reverse order by the obvious Greedy Algorithm — we select ε_i to minimize the partial sum $|\varepsilon_i r_i + \dots + \varepsilon_K r_K|$. Condition (2) insures that we never get stuck. □

Lemma 3. *Let $r_1, \dots, r_K$ be such that*

$$\sum_{|r_i| \leq c_0} |r_i| \geq \max_{1 \leq i \leq K} |r_i|. \tag{4}$$

Then there exist $\varepsilon_1, \dots, \varepsilon_K \in \{-1, +1\}$ so that (3) holds.

Proof. Placing the r_i in ascending order we see that Lemma 3 is merely a special case of Lemma 2. □

Theorem 4. *Let $A = (a_{ij})$ be a $K \times L$ matrix with all $|a_{ij}| \leq 1$. Then there exist $\delta_1, \dots, \delta_L \in \{-1, +1\}$ so that, setting*

$$r_i = \sum_{j=1}^{L} \delta_j a_{ij}, \quad 1 \leq i \leq K \tag{5}$$

(i.e. r_i is the new i-th row sum) we have

$$|r_i| \leq c_1 i^{\frac{1}{2}} \ln(i+1) \tag{6}$$

for $1 \leq i \leq K$.

(Notation: $c_0, c_1, c_2, \dots$ shall denote positive absolute constants.)

This result, which we have shown in our previous work [2] "almost" gives Theorem 1 when used with Lemma 2. Observe, however, that Theorem 4 gives no lower bound on $|r_i|$. Even if we just knew that all $|r_i| \geq 1$ (as is the case, for example, when L is odd and all $a_{ij} \in \{-1, +1\}$) then the r_i given by Theorem 4 would satisfy Lemma 2 (since $i^{\frac{1}{2}} \ln(i+1) = o(i)$) and we could make $|A| \leq c_0$. Dealing with the possibility of many small $|r_i|$ is an essential part of our paper.

A simple reduction eliminates very small lines. Suppose some line of A has the sum of the absolute values of its coefficients less than one. Delete that line and find, by induction, line shifts so that the remaining matrix has sum x with $|x| \le c_0$. At this point the line has sum r with $|r| \le 1$. Either $|x+r| \le c_0$ or $|x-r| \le c_0$ (possibly both), by selecting $c_0 \ge 1$. By either not shifting or shifting the special line $|A|$ is either $x+r$ or $x-r$ hence we insure $|A| \le c_0$.

For any $K \times L$ matrix all row sums satisfy $|r_i| \le L$. Simply by making Row Shifts we can insure $|A| \le L$. Transposing we can also insure $|A| \le K$. It therefore suffices to prove Theorem 1 when K, L are sufficiently large. This assumption shall often be made tacitly in our calculations. By symmetry we shall assume $K \ge L$. We shall show the following:

Theorem 5. *There exist K_0, c_0 such that: Let $A = (a_{ij})$ be a $K \times L$ matrix with $K \ge L \ge K_0$, all $|a_{ij}| \le 1$, and*

$$\begin{aligned} &\sum_{j=1}^{L} |a_{ij}| \ge 1, \quad 1 \le i \le K; \\ &\sum_{i=1}^{K} |a_{ij}| \ge 1, \quad 1 \le j \le L. \end{aligned} \tag{7}$$

Then there exist $\varepsilon_1, \ldots, \varepsilon_K$, $\delta_1, \ldots, \delta_L \in \{-1, +1\}$ so that

$$\left| \sum_{i=1}^{K} \sum_{j=1}^{L} \varepsilon_i \delta_j a_{ij} \right| \le c_0. \tag{8}$$

Let us return to Theorem 4. Recently Spencer [4] improved on this result by showing the upper bound $|r_i| \le c_1 i^{\frac{1}{2}}$. Spencer's theorem is the best possible apart from a constant factor. Unfortunately, Spencer's theorem does not imply any simplification in the proof of Theorem 1.

2. Probability

Let $a_1, \ldots, a_L \in [-1, +1]$ and set $\sigma = \left[\sum_{j=1}^{L} a_j^2\right]^{\frac{1}{2}}$. Let $\underline{\delta}_1, \ldots, \underline{\delta}_L$ be independent random variables with

$$\Pr(\underline{\delta}_j = 1) = \Pr(\underline{\delta}_j = -1) = \frac{1}{2}, 1 \le j \le L \tag{9}$$

and set

$$\mathbf{X} = \underline{\delta}_1 a_1 + \underline{\delta}_2 a_2 + \ldots + \underline{\delta}_L a_L. \tag{10}$$

Observe that $\mathbf{X}$ has mean zero and standard deviation σ. Proofs of the following four lemmas are given in the Appendix. Note that they would all be clearly true if $\mathbf{X}$ had a Normal Distribution.

Lemma 6. *For $\lambda > 1$,*

$$\Pr(\mathbf{X} \geq \lambda\sigma) \leq e^{-\lambda^2/2}. \tag{11}$$

Lemma 7. *There are positive absolute constants c_2, c_3 and c_4 such that: If $c_2 \leq \rho \leq \sigma$, then*

$$\frac{c_3\rho}{\sigma} \leq \Pr(\mathbf{X} \in [0, \rho]) \leq \frac{c_4\rho}{\sigma}. \tag{12}$$

Lemma 8. *There are positive absolute constants c_5, c_6, c_7, c_8, c_9 and c_{10} such that: If $\sigma < 1$, then*

$$\Pr(\mathbf{X} \in [c_5\sigma, c_6]) \geq c_7; \tag{13}$$

If $\sigma \geq 1$, then

$$\Pr(\mathbf{X} \in [c_8, c_9]) \geq \frac{c_{10}}{\sigma}. \tag{14}$$

Define $\mathbf{Y}$ by (letting $c_{11} = \max\{c_6, c_9\}$)

$$\mathbf{Y} = \begin{cases} |\mathbf{X}|, & \text{if } |X| \leq c_{11}, \\ 0, & \text{otherwise.} \end{cases}$$

Lemma 9. *If $\sigma \leq 1$, then $E(\mathbf{Y}) \geq c_{12}\sigma$; If $\sigma \geq 1$, then $E(\mathbf{Y}) \geq c_{12}/\sigma$, where E stands for expected value and c_{12} is a sufficiently small positive absolute constant.*

For convenience we define W by

$$W(\sigma) = \begin{cases} \sigma^{-1}, & \text{if } \sigma \geq 1 \\ \sigma, & \text{if } \sigma \leq 1 \end{cases}$$

and rewrite Lemma 9.

Lemma 9'. $E(\mathbf{Y}) \geq c_{12}W(\sigma)$

The following elementary result will occasionally be useful.

Lemma 10. *Let $\mathbf{Z}$ be a nonnegative random variable with $E[\mathbf{Z}] \geq \alpha$ and* $\max(\mathbf{Z}) \leq m$. *Then*

$$\Pr(\mathbf{Z} \geq \alpha/2) \geq \alpha/2m$$

Proof.

$$\alpha = E(\mathbf{Z}) \leq m\Pr(Z \geq \alpha/2) + (\alpha/2)\Pr(Z \leq \alpha/2) \leq$$
$$\leq m\Pr(Z \geq \alpha/2) + (\alpha/2).$$ □

We also shall use a discrete form of the above.

Lemma 10'. *Let $a_1, \ldots, a_k \geq 0$ with $\sum_{i=1}^{K} a_i \geq \alpha \cdot K$ and all $a_i \leq m$. Then $a_i \geq \alpha/2$ for at least $(\alpha/2m)K$ values i.*

2. The Proof

Let $A = (a_{ij})$ be a $K \times L$ matrix satisfying the conditions of Theorem 5. For $1 \leq i \leq K$ define

(16) $$\sigma_i = \left(\sum_{j=1}^{L} a_{ij}^2\right)^{\frac{1}{2}}$$

From (7) it follows that

(17) $$L^{-\frac{1}{2}} \leq \sigma_i \leq L^{\frac{1}{2}}.$$

Theorem 11. *There exist $\delta_1, \ldots, \delta_L \in \{-1, +1\}$ so that, setting*

(18) $$r_i = \sum_{j=1}^{L} \delta_j a_{ij}, \quad 1 \leq i \leq K$$

(so that r_i are the "new" row sums),

(19) $$|r_i| \leq c_{13}\sigma_i(\ln K)^{\frac{1}{2}}, \quad 1 \leq i \leq K$$

and

(20) $$\sum_{i=1}^{K} r_i^* \geq c_{14} \sum_{i=1}^{K} W(\sigma_i)$$

where we define $r_i^* = |r_i|$ if $|r_i| \leq c_{11}$ and $r_i^* = 0$, otherwise.

Proof. We consider random $\underline{\delta}_1, \ldots, \underline{\delta}_L$ as in Section 2 so that each row sum r_i is now a random variable of form (10). By Lemma 6 the probability that (19) fails for any particular i is less than $\exp[-c_{13}^2(\ln K)/2] < K^{-3}$ so the probability that (19) fails is less than K^{-2}. By Lemma 9' the term $\sum_{i=1}^{K} r_i^*$ is now a random variable with expectation at least $c_{12}\sum_{i=1}^{K} W(\sigma_i)$ and maximum value $c_{11}K$. Applying Lemma 10 and (17) (assuming $c_{14} \le c_{12}/2$)

$$\Pr((20)) \ge c_{12}\sum_{i=1}^{K}\frac{W(\sigma_i)}{2c_{11}K} \ge \frac{c_{12}}{c_{11}}\frac{\sqrt{L}^{K}}{2K} = \frac{c_{12}}{2c_{11}}L^{-\frac{1}{2}} \ge \frac{c_{12}}{2c_{11}}K^{-\frac{1}{2}}$$

which is certainly greater than $\frac{1}{K^2}$. Hence with positive probability (19) and (20) simultaneously hold — i.e. there exist specific $\delta_1, \ldots, \delta_L$ so that (19), (20) hold. □

Corollary 12. *If*

$$c_{14}\sum_{i=1}^{K} W(\sigma_i) \ge c_{13}(\max_{1\le i\le K}\sigma_i)(\ln K)^{\frac{1}{2}} \tag{21}$$

then there exist line shifts so that $|A| \le c_{11}$.

Proof. We apply the column shifts of Theorem 11 so that (19), (20) hold. If (21) holds then (4) holds with $c_0 = c_{11}$. □

Corollary 13. *If*

$$c_{14}KL^{-\frac{1}{2}} \ge c_{13}L^{\frac{1}{2}}(\ln K)^{\frac{1}{2}} \tag{22}$$

then there exist line shifts so that $|A| \le c_{11}$.

Proof. Apply (21) and the bounds (17) on the σ_i. We shall henceforth assume

$$L \le K \le \frac{c_{13}}{c_{14}}L(\ln K)^{\frac{1}{2}} \tag{23}$$

or, equivalently and more conveniently,

$$L \le K \le c_{15}L(\ln L)^{\frac{1}{2}} \tag{24}$$

Furthermore, (21) will still hold if we assume

$$\sigma_1 \ge c_{14}\sum_{i=1}^{K} W(\sigma_i)/c_{13}(\ln K)^{\frac{1}{2}} \ge c_{14}(K/\sqrt{L})/c_{13}(\ln K)^{\frac{1}{2}} \tag{25}$$

or, more conveniently

$$\sigma_1 \ge c_{16}\sqrt{L}(\ln L)^{-\frac{1}{2}}. \tag{26}$$

Now $\sigma_1^2 \ge c_{16}^2 L/(\ln L)$ is the sum of L terms a_{1j}^2, all of which are at most unity. By Lemma 10',

$$a_{1j} \ge c_{17}(\ln L)^{-\frac{1}{2}}$$

for at least

$$c_{18}L(\ln L)^{-\frac{1}{2}} \text{ values } j. \tag{27}$$

That is, A must have a "fairly large" first row. Let $\sigma_1^*, \ldots, \sigma_L^*$ denote the column standard deviations — i.e. —

$$\sigma_j^* = \Big(\sum_{i=1}^{K} a_{ij}\Big)^{\frac{1}{2}}, \quad l \le j \le L.$$

Then, by Corollary 12 applied to the transpose of A, if

$$c_{14} \sum_{j=1}^{L} W(\sigma_j^*) \ge c_{13}\big(\max_{1\le j\le L} \sigma_j^*\big)(\ln L)^{\frac{1}{2}}, \tag{28}$$

then there exist line shifts so that $|A| \le c_{11}$. As all $\sigma_j^* \le \sqrt{K}$ the right hand side is at most $c_{13}\sqrt{K}(\ln L)^{\frac{1}{2}} \le c_{13}c_{15}\sqrt{L}(\ln L)$.

Since $\sigma_j^* \ge a_{1j}$ we may rewrite (27) as

$$\sigma_j^* \ge c_{17}(\ln L)^{-\frac{1}{2}} \text{ for at least } c_{18}L(\ln L)^{-\frac{1}{2}} \text{ values } j. \tag{29}$$

Suppose at least $c_{18}L(\ln L)^{-\frac{1}{2}}/2$ of these j satisfied

$$c_{17}(\ln L)^{-\frac{1}{2}} \le \sigma_j^* \le c_{19}L^{\frac{1}{2}}(\ln L)^{-\frac{3}{2}}. \tag{30}$$

For these j's

$$W(\sigma_j^*) \ge \frac{1}{c_{19}}(\ln L)^{\frac{3}{2}} L^{-\frac{1}{2}}, \tag{31}$$

and then

$$\begin{aligned} c_{14}\sum_{j=1}^{L} W(\sigma_j^*) &\ge c_{14}(c_{18}L(\ln L)^{-\frac{1}{2}}/2)\frac{1}{c_{19}}(\ln L)^{\frac{3}{2}} L^{-\frac{1}{2}} = \\ &= \frac{c_{14}c_{18}}{2c_{19}} L^{\frac{1}{2}}(\ln L) \ge c_{13}c_{15}L^{\frac{1}{2}}(\ln L) \end{aligned} \tag{32}$$

by selecting c_{19} sufficiently small. Then (28) would be satisfied and there would exist line shifts with $|A| \le c_{11}$. Thus we may assume

$$\sigma_j^* \ge c_{19}L^{\frac{1}{2}}(\ln L)^{-\frac{3}{2}} \text{ for at least } c_{18}L(\ln L)^{-\frac{1}{2}}/2 \text{ j's}. \tag{33}$$

That is, A had many columns which were not small, as they had at least one fair sized term and for these columns to give a small value of $W(\sigma_j^*)$ they must be large.

Set $q = c_{20}(\ln L)^{\frac{7}{2}}$ and reorder so that

$$\sigma_j^* \ge c_{19}L^{\frac{1}{2}}(\ln L)^{-\frac{3}{2}} \text{ for } 1 \le j \le q. \tag{34}$$

(In fact, our argument is quite loose as we do not use the other columns guaranteed by (33).) Then (using (24))

$$\sum_{i=1}^{K}\sum_{j=1}^{q} a_{ij}^2 = \sum_{j=1}^{q}(\sigma_j^*)^2 \ge c_{20}(\ln L)^{\frac{7}{2}} c_{19}^2 L(\ln L)^{-3} \ge 2K \tag{35}$$

by setting c_{20} appropriately large. For $1 \le i \le K$ define

$$\sigma_i^{(RES)} = (\sum_{j=1}^{q} a_{ij}^2)^{\frac{1}{2}}. \tag{36}$$

Then

$$\sum_{i=1}^{K} (\sigma_i^{(RES)})^2 = \sum_{i=1}^{K} \sum_{j=1}^{q} a_{ij}^2 \ge 2K. \tag{37}$$

Since all $(\sigma_i^{(RES)})^2 \le q$ we apply Lemma 10' to deduce

$$\sigma_i^{(RES)} \ge 1 \text{ for at least } K/q \text{ values } i. \tag{38}$$

Reorder so that (38) holds for $1 \le i \le c_{21} L(\ln L)^{-\frac{7}{2}}$.

Now we apply Lemma 8. Let $\underline{\delta}_1, \ldots, \underline{\delta}_q$ be independent random variables with $\Pr(\underline{\delta}_i = +1) = \Pr(\underline{\delta}_i = -1) = 1/2$ and set

$$\mathbf{X}_i = \underline{\delta}_1 a_{i1} + \underline{\delta}_2 a_{i2} + \ldots + \underline{\delta}_q a_{iq}. \tag{39}$$

For $1 \le i \le c_{21} L(\ln L)^{-\frac{7}{2}}$,

$$\Pr(\mathbf{X}_i \in [c_8, c_9]) \ge c_{10}/\sigma_i^{(RES)} \ge c_{10} q^{-\frac{1}{2}} \ge c_{22} (\ln L)^{-\frac{7}{4}}. \tag{40}$$

Letting $\mathbf{Z}$ denote the number of $i, 1 \le i \le c_{21} L(\ln L)^{-\frac{7}{2}}$, such that $\mathbf{X}_i \in [c_8, c_9]$,

$$E(\mathbf{Z}) \ge c_{21} L(\ln L)^{-\frac{7}{2}} c_{22} (\ln L)^{-\frac{7}{4}} = c_{23} L(\ln L)^{-\frac{21}{4}}. \tag{41}$$

Therefore, there exist specific $\delta_1, \ldots, \delta_q \in \{-1, +1\}$, which we fix, so that setting

$$r_i^{(1)} = \delta_1 a_{i1} + \delta_2 a_{i2} + \ldots + \delta_q a_{iq} \tag{42}$$

and reordering the rows for convenience

$$c_8 \le r_i^{(1)} \le c_9 \text{ for } 1 \le i \le L(\ln L)^{-\frac{21}{4}} c_{23}, \tag{43}$$

$$|r_i^{(1)}| \le q = c_{20} (\ln L)^{\frac{7}{2}} \text{ for } 1 \le i \le K. \tag{44}$$

We apply Theorem 4 to the $K \times (L - q)$ matrix which remains when the first q columns of A are deleted. There exist $\delta_{q+1}, \ldots, \delta_L \in \{-1, +1\}$ so that, setting

$$r_i^{(2)} = \sum_{j=q+1}^{L} \delta_j a_{ij}, \quad 1 \le i \le K \tag{45}$$

we have

$$|r_i^{(2)}| \le c_1 i^{\frac{1}{2}} \ln(i+1), \quad 1 \le i \le K. \tag{46}$$

(Note that Theorem 4 does not require a reordering of the rows so that the $r_i^{(1)}$ retain properties (43), (44).) Now reorder the first $c_{23} L(\ln L)^{-\frac{21}{4}}$ rows in increasing order of $|r_i^{(2)}|$. Properties (43), (44), (46) are retained. Let T be that index such that

(47) $$|r_i^{(2)}| \leq 2c_9, \quad 1 \leq i \leq T$$

(48) $$|r_i^{(2)}| > 2c_9, \quad T < i \leq c_{23}L(\ln L)^{-\frac{21}{4}}.$$

(If all or none of the $|r_i^{(2)}|$ are in $[-2c_9, 2c_9]$ we may take $T = c_{23}L(\ln L)^{-\frac{21}{4}}$ or $T = 0$ respectively). For $1 \leq i \leq T$ either $|r_i^{(2)} + r_i^{(1)}| \in [c_8, 3c_9]$ or $|r_i^{(2)} - r_i^{(1)}| \in [c_8, 3c_9]$, possibly both. (The upper bound on $|r_i^{(2)} \pm r_i^{(1)}|$ follows from the upper bounds (47), (43) and holds for either sign. The lower bound, critically holds for at least one choice of sign by the lower bound (43) on $|r_i^{(1)}|$.) Let $\delta \in \{-1, +1\}$ be such that $|r_i^{(2)} + \delta r_i^{(1)}| \in [c_8, 3c_9]$ for at least half of the $i, 1 \leq i \leq T$. Once again we renumber so that these are the first $(T/2)$ indices. Then

(49) $$|r_i^{(2)} + \delta r_i^{(1)}| \in [c_8, 3c_9], \quad 1 \leq i \leq \frac{T}{2}$$

(50) $$|r_i^{(2)} + \delta r_i^{(1)}| \leq 3c_9, \quad \frac{T}{2} < i \leq T.$$

Also,

(51) $$c_9 \leq |r_i^{(2)} + \delta r_i^{(1)}| \leq c_1 i^{\frac{1}{2}} \ln(i+1) + c_9, \quad T < i \leq c_{23}L(\ln L)^{-21/4}.$$

The upper bound is the triangle inequality applied to (43), (46). The lower bound is the triangle inequality ($|x+y| \geq |x| - |y|$) applied to (48), (43). Also,

(52) $$|r_i^{(2)} + \delta r_i^{(1)}| \leq c_1 i^{1/2} \ln(i+1) + c_{20}(\ln L)^{\frac{7}{2}}, \quad c_{23}L(\ln L)^{-21/4} < i \leq K$$

by the triangle inequality applied to (44), (46).

The values $r_i = r_i^{(2)} + \delta r_i^{(1)}$ are precisely the row sums achieved after column shifts $\delta\delta_1, \ldots, \delta\delta_q, \delta_{q+1}, \ldots, \delta_L$.

Conditions (49), (50), (51), (52) insure that Lemma 2 holds for a sufficiently large constant c_0 and the Theorem 1 is proven. □

4. Appendix

Proof of Lemma 6. We have for any real number γ,

$$\Pr(\mathbf{X} \geq \gamma) = \Pr(e^{y\mathbf{X}}) \geq e^{y\gamma} \leq \frac{E(e^{y\mathbf{X}})}{e^{y\gamma}}$$

where the parameter $y > 0$ will be fixed later. Since $\mathbf{X}$ is the sum of independent random variables, we have

$$E(e^{y\mathbf{X}}) = \prod_{j=1}^{L} E(e^{y\delta_j a_j}) = \prod_{j=1}^{L}[\frac{1}{2}e^{ya_j} + \frac{1}{2}e^{-ya_j}].$$

Using the elementary inequality

$$\frac{1}{2}(e^t + e^{-t}) \leq e^{t^2/2},$$

we obtain

$$E(e^{y\mathbf{X}}) \leq e^{\frac{y^2[\Sigma_{j=1}^{L} a_j^2]}{2}} = e^{\frac{y^2\sigma^2}{2}}.$$

Hence

$$\Pr(\mathbf{X} \geq \lambda\sigma) \leq e^{\frac{1}{2}y^2\sigma^2 - y\lambda\sigma}.$$

Choosing $y = \frac{\lambda}{\sigma}$, we get

$$\Pr(\mathbf{X} \geq \lambda\sigma) \leq e^{-\lambda^2/2},$$

and Lemma 6 follows.

Proof of Lemma 7. We form all the 2^L possible sums $\pm a_1 \pm a_2 \pm \ldots \pm a_L$, and for any subset $H \subset \mathbf{R}$, denote by $Q(H)$ how many of the sums $\pm a_1 \pm a_2 \pm \ldots \pm a_L$ lie in H. The proof is based on the Fourier transform of the measure $Q(\cdot)$:

$$q(t) = \frac{1}{(2\pi)^{1/2}} \int_{\mathbf{R}} e^{-itx} dQ(x), \text{ where } i = \sqrt{-1}.$$

Observe that

$$q(t) = \frac{1}{(2\pi)^{1/2}} \sum_{2^L \, sums} e^{-it(\pm a_1 \pm a_2 \pm \ldots \pm a_L)} = \frac{1}{(2\pi)^{1/2}} \prod_{j=1}^{L} \cos(a_j t).$$

Let $\rho \in [1, \sigma]$. Let $k(t)$ be the convolution

$$k(t) = \int_{\mathbf{R}} k_0(t-u) k_0(u) du$$

of the function

$$k_0(t) = \begin{cases} \rho, & \text{if } t \in [-\frac{1}{2\rho}, \frac{1}{2\rho}] \\ 0, & \text{if } t \notin [-\frac{1}{2\rho}, \frac{1}{2\rho}] \end{cases}$$

with itself. The explicit form of $k(t)$ is $(\rho - \rho^2|t|)$ for $|t| \leq \frac{1}{\rho}$ and 0 for $|t| > \frac{1}{\rho}$. We need the following well-known general formula from Fourier analysis (see any textbook): let

$$f(t) = \int_{\mathbf{R}} g(t-u) h(u) du,$$

$$F(x) = \frac{1}{(2\pi)^{1/2}} \int_{\mathbf{R}} e^{ixt} f(t)dt,$$

$$G(x) = \frac{1}{(2\pi)^{1/2}} \int_{\mathbf{R}} e^{ixt} g(t)dt,$$

$$H(x) = \frac{1}{(2\pi)^{1/2}} \int_{\mathbf{R}} e^{ixt} h(t)dt,$$

then

$$F(x) = G(x)H(x).$$

We have

$$K_0(x) = \frac{1}{(2\pi)^{1/2}} \int_{\mathbf{R}} e^{ixt} k_0(t)dt = \frac{\rho}{(2\pi)^{1/2}} \int_{-\frac{1}{2\rho}}^{\frac{1}{2\rho}} e^{ixt}dt = \frac{\sin(\frac{x}{2\rho})}{(2\pi)^{1/2}(\frac{x}{2\rho})}.$$

Therefore, by the general formula above,

$$K(x) = \frac{1}{(2\pi)^{1/2}} \int_{\mathbf{R}} e^{ixt} k(t)dt = (K_0(x))^2 = \frac{(\sin(\frac{x}{2\rho}))^2}{2\pi(\frac{x}{2\rho})^2}.$$

Consider now Parseval's equation

$$\int_{\mathbf{R}} K(x)dQ(x) = \int_{\mathbf{R}} k(t)q(t)dt.$$

We have

$$\Pr(\mathbf{X} \in [-\rho, \rho]) = 2^{-L} \int_{-\rho}^{\rho} dQ(x) \leq 2^{-L}(2\pi)(\frac{\pi}{2})^2 \int_{\mathbf{R}} K(x)dQ(x) =$$

$$= \frac{\pi^3}{2^{L+1}} \int_{\mathbf{R}} k(t)q(t)dt = \frac{\pi^{5/2}}{2^{3/2}} \int_{-\frac{1}{\rho}}^{\frac{1}{\rho}} (\rho - \rho^2|t|) \prod_{j=1}^{L} \cos(a_j t)dt \leq$$

$$\leq \frac{\pi^{5/2}}{2^{3/2}} \rho \int_{-\frac{1}{\rho}}^{\frac{1}{\rho}} \prod_{j=1}^{L} \cos(a_j t)dt$$

Since $\cos x \leq e^{-x^2/2}$ for $|x| \leq 1$, we obtain

$$\int_{-\frac{1}{\rho}}^{\frac{1}{\rho}} \prod_{j=1}^{L} \cos(a_j t)dt \leq \int_{-\frac{1}{\rho}}^{\frac{1}{\rho}} e^{-\sigma^2 t^2/2}dt \leq \frac{c_{24}}{\sigma}.$$

Summarizing, we have for any $\rho \in [1, \sigma]$,

$$\Pr(\mathbf{X} \in [-\rho, \rho]) \leq \frac{\pi^{5/2}}{2^{3/2}} \rho \frac{c_{24}}{\sigma} = c_{25} \frac{\rho}{\sigma}.$$

This proves the upper bound in (12). Next we prove the lower bound in (12). We need the following lemma.

Lemma 14. *If $c_2 \geq 1$ is a sufficiently large absolute constant, then*

$$\int_{\mathbf{R} \setminus [-c_2\rho, c_2\rho]} K(x) dQ(x) \leq \frac{1}{2} \int_{\mathbf{R}} k(t) q(t) dt.$$

We postpone the proof of Lemma 14. We have

$$\Pr(\mathbf{X} \in [-c_2\rho, c_2\rho]) = 2^{-L} \int_{-c_2\rho}^{c_2\rho} dQ(x) \geq$$

$$\geq 2^{-L} \int_{-c_2\rho}^{c_2\rho} K(x) dQ(x) =$$

$$= 2^{-L} \int_{\mathbf{R}} k(t) q(t) dt - 2^{-L} \int_{\mathbf{R} \setminus [-c_2\rho, c_2\rho]} K(x) dQ(x).$$

It follows from Lemma 14 that for any $\rho \in [1, \sigma]$,

$$\Pr(\mathbf{X} \in [-c_2\rho, c_2\rho]) \geq 2^{-L} \cdot \frac{1}{2} \int_{\mathbf{R}} k(t) q(t) dt,$$

thus it is sufficient to show that

$$2^{-L} \int_{\mathbf{R}} k(t) q(t) dt \geq c_{26} \frac{\rho}{\sigma}.$$

But this is only a trivial calculation: using $\cos x \geq e^{-x^2}$ for $|x| \leq 1$, we have

$$2^{-L} \int_{\mathbf{R}} k(t) q(t) dt = \frac{1}{(2\pi)^{1/2}} \int_{-\frac{1}{\rho}}^{\frac{1}{\rho}} (\rho - \rho^2 |t|) \prod_{j=1}^{L} \cos(a_j t) dt \geq$$

$$\geq \frac{1}{(2\pi)^{1/2}} \int_{-\frac{1}{\rho}}^{\frac{1}{\rho}} (\rho - \rho^2 |t|) e^{-\sigma^2 t^2} dt \geq c_{26} \frac{\rho}{\sigma}$$

as required. It remains to prove Lemma 14.

Proof of Lemma 14. First we show that

$$\sup_{y\in\mathbf{R}} \int_{-\rho+y}^{\rho+y} dQ(x) \le \frac{\pi^3}{2} \int_{\mathbf{R}} k(t)q(t)dt. \tag{53}$$

Since

$$K(x-y) = \frac{1}{(2\pi)^{1/2}} \int_{\mathbf{R}} e^{i(x-y)t} k(t)dt = \frac{1}{(2\pi)^{1/2}} \int_{\mathbf{R}} e^{ixt} k(t) e^{-iyt} dt,$$

it follows that $K(x-y)$ (as a function of x) is the Fourier inverse of $k(t)e^{-iyt}$. Thus by Parseval's equation

$$\int_{\mathbf{R}} K(x)dQ(x+y) = \int_{\mathbf{R}} K(x-y)dQ(x) =$$

$$= \int_{\mathbf{R}} k(t)e^{-iyt} q(t)dt \le \int_{\mathbf{R}} |k(t)q(t)|dt =$$

$$= \int_{\mathbf{R}} k(t)q(t)dt$$

since $k(t) = 0$ for $|t| > \frac{1}{\rho}$. Therefore,

$$\int_{-\rho+y}^{\rho+y} dQ(x) \le 2\pi(\frac{\pi}{2})^2 \cdot \int_{\mathbf{R}} K(x)dQ(x+y) \le \frac{\pi^3}{2} \int_{\mathbf{R}} k(t)q(t)dt,$$

and (53) is verified. We have (assuming that c_2 is an odd integer)

$$\int_{\mathbf{R}\setminus[-c_2\rho, c_2\rho]} K(x)dQ(x) = \sum_{\substack{n\in Z \\ 2n-1\ge c_2 \\ or\ 2n+1\le -c_2}} \int_{(2n-1)\rho}^{(2n+1)\rho} K(x)dQ(x) \le$$

$$\le \sum_{\substack{n\in Z \\ 2n-1\ge c_2 \\ or\ 2n+1\le -c_2}} \left(\max_{x\in[(2n-1)\rho,(2n+1)\rho]} K(x)\right) \int_{(2n-1)\rho}^{(2n+1)\rho} dQ(x).$$

Elementary calculation shows that for any $n \in \mathbf{Z}$ and $x \in [(2n-1)\rho, (2n+1)\rho]$

$$K(x) = \frac{1}{(2\pi)} \left(\frac{\sin(\frac{x}{2\rho})}{(\frac{x}{2\rho})}\right)^2 \le \frac{c_{27}}{1+n^2}.$$

Therefore, by (53) we have

$$\int\limits_{\mathbf{R}\setminus[-c_2\rho,c_2\rho]} K(x)dQ(x) \le$$

$$\le \frac{\pi^3}{2}\left(\sum_{\substack{n\in Z:\\ 2n-1\ge c_2\\ or 2n+1\le -c_2}} \frac{c_{27}}{1+n^2}\right)\left(\int\limits_{\mathbf{R}} k(t)q(t)dt\right) \le$$

$$\le \frac{1}{2}\int\limits_{\mathbf{R}} k(t)q(t)dt$$

if c_2 is sufficiently large. Lemma 14 follows. This completes the proof of Lemma 7. □

Proof of Lemma 8. We have

$$\int_0^\infty \Pr(\mathbf{X}^2 \ge y)dy = E(\mathbf{X}^2) = \sigma^2.$$

Clearly

$$\int_0^{\frac{\sigma^2}{4}} \Pr(\mathbf{X}^2 \ge y)dy \le \frac{\sigma^2}{4}.$$

On the other hand, by Lemma 6 we have

$$\int_t^\infty \Pr(\mathbf{X}^2 \ge y)dy \le 2\int_t^\infty e^{-y/2\sigma^2}dy \le \frac{\sigma^2}{4}$$

provided $t = c_{28}\sigma^2$ and c_{28} is sufficiently large. Summarizing,

$$\int_{\frac{\sigma^2}{4}}^{c_{28}\sigma^2} \Pr(\mathbf{X}^2 \ge y)dy \ge \sigma^2 - \frac{\sigma^2}{4} - \frac{\sigma^2}{4} = \frac{\sigma^2}{2}.$$

It follows that

$$\Pr(|\mathbf{X}| \ge \frac{\sigma}{2}) \ge \frac{\frac{\sigma^2}{2}}{c_{28}\sigma^2} = \frac{1}{2c_{28}}.$$

Again by Lemma 6, $\Pr(|\mathbf{X}| \ge c_{29}\sigma) \le \frac{1}{4c_{28}}$ provided c_{29} is sufficiently large. Hence

$$\Pr(|\mathbf{X}| \in [\frac{\sigma}{2}, c_{29}\sigma]) \ge \frac{1}{2c_{29}} - \frac{1}{4c_{29}} = \frac{1}{4c_{29}}, \tag{54}$$

and (13) follows. Next we prove (14). By (54) we can assume that $\sigma \geq \frac{2c_2 \cdot c_4}{c_3}$. Let $c_8 = c_2$ and $c_9 = \frac{2c_2c_4}{c_3}$, then by Lemma 7 we conclude that

$$\Pr(\mathbf{X} \in [c_8, c_9]) = \Pr(\mathbf{X} \in [0, c_9]) - \Pr(\mathbf{X} \in [0, c_8]) \geq c_3 \frac{c_9}{\sigma} - c_4 \frac{c_8}{\sigma} =$$
$$= \frac{c_2 c_4}{\sigma} = \frac{c_{10}}{\sigma}$$

and (14) follows. Lemma 8 is proved.

Proof of Lemma 9. It follows immediately from Lemmas 7-8.

References

[1] J.Beck and J.Spencer, Balancing matrices with line shifts, *Combinatorica* **3** (1983), 299-304.

[2] J.Beck and J.Spencer, Integral approximation sequences,*Math. Programming* **30** (1984), 88-98.

[3] J.Komlós and M.Sulyok, On the sum of elements of ±1 matrices, *in: Combinatorial Theory and Its Applications (Erdős et al.,eds.), North-Holland* 1970, 721-728.

[4] J.Spencer, Six standard deviations suffice, *Trans. Amer. Math. Soc.* **289** (1985), 679-706.

J. Beck
L. Eötvös University
Budapest

J. Spencer
SUNY at Stony Brook
Stony Brook, N.Y. 11794 U.S.A

3. A Few Remarks on Orientation of Graphs and Ramsey Theory

M. Cochand and P. Duchet

Abstract

Let $\mathcal{D}$ denote a class of digraph such that every induced digraph in $\mathcal{D}$ is in $\mathcal{D}$ again. Then either $\mathcal{D}$ contains all acyclic digraphs or almost no graph has an orientation in $\mathcal{D}$. Proofs and variations on this theme are discussed. Some open problems in Ramsey theory are raised.

I. Introduction

This paper deals with finite graphs. A *geodesic* from a to b in a directed graph is a shortest directed path from a to b. R.C. Entringer conjectured an affirmative answer to the following problem:

(1.1) *Does every graph have an orientation in which every geodesic is unique?*

A similar question was raised by Bienia and Meyniel [3]:

(1.2) *Does every graph have an orientation in which, for every pair a, b of nodes, all induced paths from a to b have the same parity?*

Bollobás [4] showed that large Paley graphs are counterexamples for (1.1) and proved that almost no graph has an orientation with unique geodesics. We show here that a simple application of results from Ramsey theory suffices to prove that almost all graphs are counterexamples to both problems (1.1) and (1.2). We give in section 2 a simple new constructive proof (which may be of independent interest) of the following result due to Rödl:

Theorem A. [**Rödl [14]**]. *Given an acyclic digraph D, there exists an undirected graph G whose every acyclic orientation contains an induced copy of D.*

Notice that the conclusion of theorem A is no longer true if D is not acyclic (i.e. if D contains a directed cycle). If D has n vertices, our proof yields a graph G whose order is an exponential iterated n times, but this bound can be probably considerably reduced (see Problem 2.4 and Conjecture 2.6) The following stronger form of Theorem A seems not to have been stated previously:

Theorem B. *Given an acyclic digraph D, there exists an undirected graph G whose every orientation contains an induced copy of D.*

The insertion of Theorems A and B in Ramsey theory is discussed in Section 3.

For instance we point out that Deuber's proof of the induced version of Ramsey theorem [5, 6, 15] actually proves Theorem B and by the same way implies the (apparently stronger) induced versions of Ramsey theorem for ordered graphs and for acyclic graphs [8, 11, 13, 14]. A conjecture is stated for arbitrary digraphs.

The result mentioned in our abstract is proved in Section 4, where small explicit counterexamples to problems (1.1) and (1.2) are also presented. These are obtained by an application of our construction for Theorem A.

2. A Construction for Theorem A

In what follows $[c]$ denotes the set of naturals $1, \ldots, c$. Set difference is denoted by $\setminus$ and symmetric difference by $\triangle$. By $\max(X)$ we mean the largest element of a finite non emptyset $X \subset \mathbf{N}$.

Lemma 2.1. *For every mapping $\varepsilon : 2^{[c]} \to [c]$ there exist different subsets $A, B \in 2^{[c]}$ such that $\varepsilon(A) = \varepsilon(B) = \max(A \triangle B)$.*

Proof. The lemma is obvious for $c = 1$. Supposing $c > 1$, we proceed by induction on c. If $\varepsilon(X) < c$ for every $X \in 2^{[c]}$, then, by the induction hypothesis, the lemma is true with the restriction of ε to the set $2^{[c-1]}$; the conclusion follows. Otherwise $f(A) = c$ for some $A \in 2^{[c]}$. Set $\mathcal{X} = \{X \in 2^{[c]}; c \in A \triangle X\}$. If $f(B) = c$ for some $B \in \mathcal{X}$, the pair A, B has the required property. Otherwise every member of $\mathcal{X}$ is uniquely determined by its trace on $[c-1]$; hence, applying the induction hypothesis to the mapping $\varepsilon' : 2^{[c-1]} \to [c-1]$ defined by $\varepsilon'(X \cap [c-1]) = \varepsilon(X)$ for $X \in \mathcal{X}$, we are done. □

Proof of theorem A. By induction on n, the number of D-vertices: The theorem is obvious for $n = 1$, so we assume $n > 1$ and we suppose the theorem holds for all digraphs with less than n vertices if $D = \emptyset$. Let s be a sink of D and denote by S the set of predecessors of s in D. By our hypothesis, there exists an undirected graph Γ any acyclic orientation of which contains an induced copy of $D \setminus s$. If $S = \emptyset$, the graph Γ augmented with an isolated vertex satisfies the graph Γ augmented with an isolated vertex satisfies the requirements of, so we may assume $S \neq \emptyset$.

Let us consider a family $\mathcal{C}$ of "unavoidable copies" of D and S in Γ: formally $\mathcal{C}$ is a family of c pairs $(V_i, S_i)_{1 \leq i \leq c}$ the following properties:

(1) Each V_i is a set of Γ-vertices.
(2) For $i \in [c]$, we have $S_i \subseteq V_i$.
(3) For any acyclic orientation Γ^ε of Γ, there exists $i \in [c]$ and there exists an isomorphism φ_i from D onto the subdigraph of Γ^ε induced by V_i, so that φ_i maps S to S_i.

To cut it short, property (3) is abridged in "integer i is *active* in Γ^ε". To construct the graph G we take 2^c disjoint copies of Γ and we add new edges as follows: indexing the copies of Γ by subsets of $[c]$ and denoting by $\Gamma(X)$, $V_i(X)$, $S_i(X)$ respectively the copy of Γ, V_i, S_i that corresponds to $X \in 2^{[c]}$ (for each $i \in [c]$), we join by additional edges all vertices of $S_i(A)$ to all vertices of $S_i(B)$ every time that $i = \max(A \triangle B)$, with $A, B \in 2^{[c]}$.

To show G has the required property, let us consider an acircuitic orientation G^ε of G.

For each $X \in 2^{[c]}$, the induced subgraph $\Gamma(X)$ receives an induced acircuitic orientation $\Gamma^\varepsilon(X)$ for which we can choose an active integer $\varepsilon(X) \in [c]$. Lemma 2.1 applies: there exist $A, B \in 2^{[c]}$ such that $\varepsilon(A) = \varepsilon(B) = \max(A \triangle B)$. By construction all edges between $S_{\varepsilon(A)}(A)$ and $S_{\varepsilon(B)}(B)$ are present in Γ; since Γ^ε has no circuit and A, B play the same role, we may suppose that some vertex b of $S_{\varepsilon(B)}(B)$ has no successor in $S_{\varepsilon(A)}(A)$. Thus the set $V_{\varepsilon(A)}(A) \cup \{b\}$ induces a copy of D in G^ε. This achieves the proof of Theorem A. □

Remark 2.2. There is no possible interference between disjoint unavoidable copies of D in Γ; thus the number c of unavoidable copies of (D, S) in Γ can be replaced in the construction by the integer c' which is the chromatic index of $\mathcal{C}$: formally $\mathcal{C}$ is replaced by a family $\mathcal{C}' = (C_\alpha)_{1 \leq \alpha \leq c'}$ of $[c]$-subsets such that $V_i \cap V_j = \emptyset$ for every $i, j \in c_\alpha$, $i \neq j$. In G, vertices of $S_i(A)$ are joined to vertices of $S_j(B)$ every time that $i, j \in c_\alpha$ with $\alpha = \max(A \triangle B)$.

A graph $H = (V, E)$ is said to be *k-emulsive* if it admits a k-edge-labelling $\psi : E \to [k]$ with the following property: for any k-vertex-labelling $\varphi : V \to [k]$, there exists an edges $e = \{x, y\}$ in H such that $\varphi(x) = \varphi(y) = \psi(e)$. Lemma 2.1. shows that the complete graph with 2^k vertices is k-emulsive. Obviously, a k-emulsive graph has no proper k-edge-coloring, hence has maximum degree greater than k. By remark 2.2, what is important for the size of the graph constructed in the proof of Theorem A is the chromatic index of a k-emulsive graph. This observation motivates the following problem:

Problem 2.3. Let $d(k)$ denote the minimal value of the maximum degree of a k-emulsive graph. Does $d(k)$ have a polynomial growth?

We were unable to decide whether $d(k)$ is $0(k)$ or not.

Remark 2.4. Construction, when applied successively to pendant vertices of a given directed tree T yields actually a graph whose every orientation contains an induced copy of T; Theorem B for directed trees follows.

Conjecture 2.5. For some absolute constant $C > 0$, there exists a graph G with C^n vertices such that any orientation of G contains an induced copy of any directed tree with n vertices.

3. Variations

In this section, we show how Theorem A and B are related to other Ramsey type results. Theorem A appears as Theorem 3.3c of [14], when properly stated under the following form:

Theorem 3.1. **[Rödl [14]].** *For any positive integer n, there exists a graph R such that any acyclic digraph with n vertices is an induced subdigraph of every acyclic orientation of R.* □

As a matter of fact, what is proved in [14] is the following:

Theorem 3.2. **[Rödl [14]].** *For any acyclic digraph D, there exists a graph R such that any acyclic orientation R^a of R contains, for every bicolouring of R^a-arcs, a monochrome induced copy of D.* □

It is remarkable that Theorem B can be readily derived from theorem 3.2.

Proof of theorem B via theorem 3.2. Let D be an acyclic digraph and denote by D' the disjoint union of D and its reversal D^{-1}. Let R denote a graph that satisfies the requirements of theorem 3.2, when applied to D' (instead of D). We claim that any orientation R^ε of R contains an induced copy of D. Indeed, choose an acyclic orientation R^a of R and color blue (respectively red) the R^a-arcs which are present in R^ε (respectively, whose reversals are present in R^ε).

By theorem 3.2, acyclic digraph R^a contains some monochrome induced copy of $D \cup D^{-1}$, hence digraph R^ϵ also contains an induced copy of $D \cup D^{-1}$ and a fortiori of D. □

With similar tricks, one can derive Theorem B from ordered versions of theorems 3.1 and 3.2, where the expressions "acyclic digraph" and "acyclic orientation" are respectively replaced by "ordered graph" and "ordering of vertices". For details on these variants and generalizations see [15, 11, 12, 13]. It is worthy to note that these ordered versions can easily be derived again from theorem 3.2: Orienting the edges of any ordered graph $(H, \leq)$ in the increasing direction (with respect to $\leq$), we obtain an acyclic digraph H^a. We embed H^a in a larger acyclic digraph D by adding to H^a a new vertex v_{xy} and two arcs (x, v_{xy}) and (v_{xy}, y) each time H^a contains two non adjacent vertices x, y such that y follows x (with respect to the linear ordering $\leq$). Digraph H^a is an induced subdigraph of D and its ordering $\leq$ is uniquely determined by D (since D has an Hamiltonian path). Hence, applying theorem 3.2 to D, we have a graph R that contains a monotone and monochrome induced copy of $(H, \leq)$, for every ordering of R-vertices and for every bicolouring of R-edges.

At last, another simple derivation of Theorem B deserves mention. The induced version of Ramsey theorem may be stated as follows:

Theorem 3.3. [**(Deuber [5]; Erdős, Hajnal, Pósa [6]; Rödl [15]]**. *For any pair of graphs (G_1, G_2) there exists a graph R with the following property: for any bicolouring of R-edge, say with red and blue, the graph R contains either a red induced copy of G_1 or a blue copy of G_2.* □

A short proof of theorem 3.3 can be found in [10]. Deuber's proof is rather involved (see [7] for a sketch) but has a great advantage: it provides also, with minor adaptation, a proof of Theorem B. Not entering into details, we only mention here the nature of the modification: Deuber's proof use double induction and assumes the validity of the theorem for pairs (G'_1, G_2) and (G_1, G'_2) where $G'_i (i = 1, 2)$ is of the form $G_i \setminus v_i$. Dealing with a pair (D_1, D_2) of digraphs, we apply Deuber's construction to the underlying undirected graphs, choosing v_i as a sink of $D_i (i = 1, 2)$. We obtain by this way a graph R. Any orientation of R will contain either an induced copy of D_1 or an induced copy of D_2. The reason is that an exact translation of Deuber's argument in term of orientation is possible: red and blue edges become respectively arcs oriented forwards and arcs oriented backwards, when referred to the order in which vertices are constructed.

Consequently, Deuber's construction in germ an independent proof of Theorem 3.2: indeed, Theorem 3.3 and B together trivially imply the following statement which can be considered as the standard theorem for edge partitions:

Theorem 3.4. *Given any acyclic digraph D, there exists a graph R such that, for every bicolouring of R-edges, any orientation of R contains a monochrome induced copy of D.* □

The assumption "acyclic" is of course essential, but the following question is (as far as we know) unsolved:

Conjecture 3.5. For any digraph D, there exists a graph whose every strongly connected orientation contains an induced copy of D.

The answer is affirmative for acyclic graphs, directed cycles and tournaments.

4. Applications

Problems (1.1) and (1.2), and similar problems relative to orientation of graphs are answered by Theorem A, B or their following corollary:

Theorem C. *Let $\mathcal{D}$ be a class of digraphs such that every induced subdigraph of a member of $\mathcal{D}$ is again in $\mathcal{D}$. Then either all acyclic digraphs are $\mathcal{D}$ or almost no graph admits an orientation in $\mathcal{D}$.*

Proof. Suppose D is an acyclic digraph not in $\mathcal{D}$; by theorem B, every orientation of some graph R contains an induced copy of D, R has no orientation in $\mathcal{D}$. Almost every graph contains R as an induced subgraph, hence has no orientation in $\mathcal{D}$. □

Let $D(p,q)$ denote the acyclic digraph with $p+q$ vertices consisting of two disjoint directed paths of respective lengths p and q, with common endpoints. Applying Theorem B to $D(2,2)$ (the "diamond") and to $D(3,2)$ (the "skew diamond") respectively, we obtain counterexamples required for problems (1.1) and (1.2) respectively. Of course these counterexamples are very large. Smaller ones can be constructed, as indicated below.

Proposition 4.2. *Every orientation of the direct product $C_5 \cdot \overline{K}_3$ contains an induced diamond.*

Proof. The verification is left to the reader; we recall that $C_5 \cdot \overline{K}_3$ has 15 vertices v_{ij} and 45 edges of the form $\{v_{ij}, v_{(i+1)j_k}\}$ where $i \in \mathbf{Z}/5\mathbf{Z}, j_k \in \mathbf{Z}/3\mathbf{Z}$. □

Proposition 4.3. *There is a graph with 2560 vertices whose every orientation contains a skew diamond.*

Proof (Sketched). Let T denote the directed tree obtained from the skew diamond by deletion of its unique sink. Choose as graph Γ the graph on 10 vertices formed with two disjoint 5-cycles joined by an edge. Every orientation of Γ contains an induced copy of T; more precisely, using the terminology of (2.1), we have a list $\mathcal{C}$ of 14 "unavoidable copies" of T in Γ. Applying the construction of (2.1) and taking into account Remark 2.2, we obtain (since $\mathcal{C}$ has chromatic index 8) a graph G with 10.2^8 vertices such that any *acyclic* orientation of G contains an induced skew diamond. It turns out that actually *every orientation* of G contains an induced skew diamond: observe that the digraph obtained from the skew diamond by reversing the third arc of the 3-path is again a skew diamond. □

Remark 4.4. The above observation holds for every digraph of the form $D(p+1,p)$, hence our construction (2.1) suffices to establish Theorem B for those digraphs.

To conclude, let us mention that Theorem A provides graphs which are not the covering diagram (="Hasse diagram") of any partial order (a problem solved by Ore). Recently, Nešetřil and Rödl proved the existence of such graphs with arbitrarily large girth [12,13].

Acknowledgements. Our sincere thanks are due to P. Hell for pointing out to us a quick proof of Theorem B via the induced version of Ramsey theorem for ordered graphs and/or acyclic relations [14,15].

References

[1] *Infinite and Finite Sets*, Coll. Math. Soc. János Bolyai **10** (Keszthely 1973), Budapest 1976.

[2] *Combinatorics*, Coll. Math. Soc. János Bolyai **18** (Keszthely 1976), North Holland Pub. Co., 1978.

[3] V. Bienia, H. Meyniel, *Séminaire du Lundi*, MSH, Paris, 1984.

[4] B. Bollobás, Geodesics in oriented graphs, *Ann. Disc. Math.* **20** (1984) 76-73.

[5] W. Deuber, Generalizations of Ramsey's theorem, in *Infinite and Finite Sets*, Coll. Math. Soc. János Bolyai **10** (Keszthely 1973), Budapest 1976 ([1]) 323-332.

[6] P. Erdős, A. Hajnal, L. Pósa, Strong embeddings of graphs into colored graphs, in *Infinite and Finite Sets*, Coll. Math. Soc. János Bolyai **10** (Keszthely 1973), Budapest 1976. ([1]) 585-595.

[7] R. L. Graham, B.L. Rothschild, Some recent developments in Ramsey theory: in *Combinatorics*, M. Hall Jr. and J.H. Van Lint eds., D. Reidel Publ. Co., Dordrecht-Boston 1975, 261-276.

[8] J. Nešetřil, V.Rödl, Partition (Ramsey) Theory - A survey, in *Combinatorics*, Coll. Math. Soc. János Bolyai **18** (Keszthely 1976), North Holland Pub. Co., 1978. ([2]) 759-792.

[9] J. Nešetřil, V. Rödl, Partitions of finite relational and set systems, *J. Comb. Th. (A)*, **22** (1977) 289-312.

[10] J. Nešetřil, V. Rödl, A simple proof of the Galvin-Ramsey property of the class of all finite graphs and a dimension of a graph, *Disc. Math.* **23** (1978) 49-55.

[11] J. Nešetřil, V. Rödl, Extensions of full Ramsey theorem in the categories of all graphs and the categories of relations, Pamphlet 2, 1974 (mimeo Charles Univ., Praha).

[12] J. Nešetřil, V. Rödl, On a probabilistic graph-theoretical method, *Proc. Amer. Math. Soc.* **72** (1978) 417-421.

[13] J. Nešetřil, V. Rödl, Combinatorial partitions of finite posets and lattices — Ramsey lattices *Alg. Univ.* **19** (1984) 106-119.

[14] V. Rödl, A generalization of Ramsey theorem, in: *Graphs, Hypergraphs and Block Systems* (Zielona Gora 1976) 211-220.

[15] V. Rödl, Dimension of a graph and a generalization of Ramsey theorem (Czech), *Thesis*, Charles Univ., Praha (1973).

M. Cochand
E.P.L., Lausanne

P. Duchet
C.N.R.S., Paris

4. On a Conjecture of Roth and Some Related Problems I

P. Erdős, A. Sárközy[1], V.T. Sós[2]

1. Introduction

Let $\mathcal{N}$ denote the set of positive integers and put $[1,N] = \{1,\dots,N\}$. We use $\mid S \mid$ to denote the cardinality of the finite set S. If S is a given set and $\mathcal{A}_1,\dots,\mathcal{A}_k$ are subsets of S with

$$S = \cup_{i=1}^{k} \mathcal{A}_i, \quad \mathcal{A}_i \cap \mathcal{A}_j = \emptyset \quad \text{for} \quad i \neq j,$$

then $\{\mathcal{A}_1,\dots,\mathcal{A}_k\}$ will be called a k-partition (or k-colouring) of S, and the subsets $\mathcal{A}_1,\dots,\mathcal{A}_k$ will be referred to as classes. Let $f : \mathcal{N}^t \to \mathcal{N}$ be a given function. If

$$n = f(a_1,\dots,a_t) \tag{1}$$

with $a_1,\dots,a_t$ belonging to the same class, then this will be called a *monochromatic* representation of n in the form (1)

For a fixed k-partition and f we consider the set of integers, which have a monochromatic representation and investigate

a) how dense this set must be?
b) for which $S \subseteq \mathcal{N}$ it must contain an element in S?
c) what sort of structural properties this set has?

We consider first the case $f(x_1,x_2) = x_1 + x_2$.
Let C resp. C^2 denote the set of integers resp. the set of even integers which have a monochromatic representation in the form

$$n = a_1 + a_2 \quad \text{with} \quad a_1 \neq a_2 \tag{2}$$

Put $C_M = C \cap [1,M]$ and $C^2_M = C^2 \cap [1,M]$.

---

(1),(2) Research partially supported by Hungarian National Foundation for Scientific Research grant no. 1811

K.F. Roth conjectured [see [4] and [9], p.112) that there is an absolute constant $c > O$ such that for an arbitrary k-partition

$$| \mathrm{C}_M | > cM. \tag{3}$$

(Note that if also $a_1 = a_2$ is allowed, then this is trivial.)

We prove this conjecture in a sharper and more general form. We study some related problems too.

The Case $f(x_1, x_2) = x_1 + x_2$

Theorem 1.

(i) *To every $k \geq 2$ there exists an $M_0(k)$ such that for an arbitrary k-partition of $\mathcal{N}$*

$$| \mathrm{C}_M^2 | > \frac{M}{2} - 3M^{1-2^{-k-1}} \quad \text{if} \quad M > M_0(k). \tag{4}$$

Moreover

(ii) *For every 2-partition*

$$| \mathrm{C}_M^2 | > \frac{M}{2} - \left(\log\left(\frac{1+\sqrt{5}}{2}\right)\right)^{-1} \log M \tag{5}$$

(iii) *There is a 2-partition so that*

$$2^n \notin \mathrm{C}^2 \quad \text{for} \quad n \in \mathcal{N} \tag{6}$$

Proof.

(i) The proof will be based on the following

Lemma 1. *If $d \in \mathcal{N}$, $M > M_0(d)$, $\mathcal{B} \subseteq [1, M]$ and*

$$| \mathcal{B} | > 3M^{1-2^{-d}} \tag{7}$$

then there exist positive integers u, $v_1, \ldots, v_d$ such that $v_i \neq v_j$ for $i \neq j$ and all the 2^d sums

$$u + \sum_{i=1}^{d} \varepsilon_i v_i, \quad \varepsilon_i \in \{0, 1\} \tag{8}$$

belong to $\mathcal{B}$.

This is a density version of Hilbert's lemma [10] (which is considered as the first Ramsey-type result). See also [8]. It can be proved similarly to Lemma 7 in [14] (see also [3] and [20]). However for the sake of completeness, we give the proof here.

Proof of Lemma 1. It suffices to show the existence of sets $\mathcal{B}_0, \mathcal{B}_1, \ldots, \mathcal{B}_d$ and distinct positive integers $v_1, v_2, \ldots, v_d$ such that

$$\mathcal{B}_0 = \mathcal{B}, \tag{9}$$

$$\mathcal{B}_j \cup \{b + v_j : b \in \mathcal{B}_j\} \subset \mathcal{B}_{j-1} \quad \text{for} \quad j = 1, 2, \ldots, d \tag{10}$$

and

$$|\mathcal{B}_j| \geq |\mathcal{B}|^{2^j} (3M)^{-(2^j-1)} \quad \text{for} \quad j = 0, 1, 2, \ldots, d. \tag{11}$$

In fact, if $\mathcal{B}_0, \mathcal{B}_1, \ldots, \mathcal{B}_d$, $v_1, \ldots, v_d$ satisfy these conditions and $u \in \mathcal{B}_d$, then by (9) and (10), $u + \sum_{i=1}^{d} \varepsilon_i v_i \in \mathcal{B}$ for $\varepsilon_i = 0$ or 1, while (7) and (11) imply that $\mathcal{B}_d$ is not empty. This then will complete the proof of Lemma 1.

We are going to construct $\mathcal{B}_0, \mathcal{B}_1 \ldots, \mathcal{B}_d$, $v_1, \ldots, v_d$ recursively. Let $\mathcal{B}_0 = \mathcal{B}$. Assume now that $0 \leq j \leq d-1$ and, in the case $j > 0$, $v_1, \ldots, v_j$ have already been defined. For $1 \leq h \leq M-1$, let $f(\mathcal{B}_j, h)$ denote the number of solutions of

$$b - b' = h, \quad \text{where} \quad b, b' \in \mathcal{B}_j.$$

Then in order to define $\mathcal{B}_{j+1}$ and v_{j+1}, we need an estimate for

$$L = \max f(\mathcal{B}_j, h)$$

where the maximum is over all h with $h \in [1, M]$, $h \notin \{v_1, v_2, \ldots, v_j\}$.

Clearly, for all h we have $f(\mathcal{B}_j, h) \leq |\mathcal{B}_j|$. Also

$$\sum_{h=1}^{M-1} f(\mathcal{B}_j, h) = \binom{|\mathcal{B}_j|}{2} \tag{12}$$

since $b - b' \in [1, M]$ for any pair $b, b' \in \mathcal{B}_j$ with $b > b'$. If we majorize $f(\mathcal{B}_j, h)$ by $|\mathcal{B}_j|$ for $h \in \{v_1, v_2, \ldots, v_j\}$ and by L otherwise, (12) implies

$$\binom{|\mathcal{B}_j|}{2} \leq j\,|\mathcal{B}_j| + (M-1-j)L \leq j\,|\mathcal{B}_j| + LM,$$

so that

$$L > \frac{1}{2M}\left(|\mathcal{B}_j|^2 - |\mathcal{B}_j| - 2j\,|\mathcal{B}_j|\right) = \frac{|\mathcal{B}_j|}{3M}\left(\frac{3}{2}\,|\mathcal{B}_j| - \frac{3}{2} - 3j\right). \tag{13}$$

From (7) and (11), we have (for M larger than some absolute and computable constant)

$$| \mathcal{B}_j | \geq | \mathcal{B} |^{2^j} (3M)^{-(2^j-1)} > \left(3M^{1-2^{-d}}\right)^{2^j} (3M)^{-(2^j-1)} =$$

$$= 3M^{1-2^{j-d}} \geq 3M^{1-2^{-1}} > 3+6d > 3+6j,$$

so that (11) and (13) imply

$$L > \frac{| \mathcal{B}_j |}{3M} \cdot | \mathcal{B}_j | \geq \frac{1}{3M} \left(| \mathcal{B} |^{2^j} (3M)^{(2^j-1)} \right)^2 =$$

(14)

$$= | \mathcal{B} |^{2^{j+1}} (3M)^{-(+2^{j+1}-1)}.$$

Let $v_{j+1} \in [1, M] \backslash \{v_1, v_2, \ldots, v_j\}$ denote an integer for which the maximum in the definition of L is attained, i.e., $L = f(\mathcal{B}_j, v_{j+1})$ with $v_{j+1} \notin \{v_1, v_2, \ldots, v_j\}$, and let

$$\mathcal{B}_{j+1} = \{b : b \in \mathcal{B}_j,\ b + v_{j+1} \in \mathcal{B}_j\}.$$

Thus (10) holds for $j+1$ in place of j and since $| \mathcal{B}_{j+1} | = L$, (14) implies that (11) holds also for $j+1$ in place of j. This completes the proof of the existence of $\mathcal{B}_0, \mathcal{B}_1, \ldots, \mathcal{B}_d,\ v_1, \ldots, v_d$ with the desired properties, so that Lemma 1 is proved.

To prove the first statement in Theorem 1, we assume that there are more than $3M^{1-2^{-k-1}}$ even integers not exceeding M which do not have a monochromatic representation in the form (2); let us denote the set of these integers by $\mathcal{B}$. Then (3) holds with $k+1$ in place of d, thus if M is sufficiently large, then by Lemma 1 there exist positive integers $u, v_1, v_2, \ldots, v_{k+1}$ such that all the sums

$$u + \sum_{i=1}^{k+1} \varepsilon_i v_i \quad \text{where} \quad \varepsilon_i = 0 \quad \text{or} \quad 1$$

belong to $\mathcal{B}$. Then

$$u = u + \sum_{i=1}^{k+1} 0 \cdot v_i \in \mathcal{B}$$

and since $\mathcal{B}$ consists of even numbers, thus also $u = 2z$ is even. The integers $z+v_1, z+v_2, \ldots, z+v_{k+1}$ are distinct, thus by the pigeon hole principle, there exist $1 \leq i < j \leq k+1$ such that $a_1 = z + v_i$ and $a_2 = z + v_j$ belong to the same class. Then $a_1 + a_2$ is a monochromatic sum with $a_1 \neq a_2$, and

$$a_1 + a_2 = (z + v_i) + (z + v_j) = 2z + v_i + v_j = u + v_i + v_j$$

But this contradicts the definition of $\mathcal{B}$, and the proof of the first half of Theorem 1 is completed.

(ii) Let $\mathcal{B} = \{b_1, b_2, \ldots, b_t\}$ (where $b_1 < b_2 < \ldots < b_t$) denote the set of those even integers not exceeding $2M$ which do not have a monochromatic representation in the form (2).
Suppose

$$b_{j+2} < b_j + b_{j+1} \tag{15}$$

for some j. Then there are positive integers x, y, z for which

$$x + y = b_j$$
$$x + z = b_{j+1}$$
$$y + z = b_{j+2}$$

At least two of these numbers belong to the same class. This contradicts to the definition of $\mathcal{B}$. Hence for every j

$$b_{j+2} \geq b_j + b_{j+1} \tag{16}$$

which proves (ii)

To prove (iii) we define the set $\mathcal{A}_1$ recursively. Let $1 \in \mathcal{A}_1$. If $\mathcal{A} \cap [1, 2^{k-1}]$ has been defined, then let $2^k \in \mathcal{A}_1$ and for $2^{k-1} < n < 2^k$, $n \in \mathcal{A}_1$ iff $2^k - n \notin \mathcal{A}_1 \cap [1, 2^{k-1}]$. Furthermore let $\mathcal{A}_2 = \mathcal{N} \backslash \mathcal{A}_1$. Then obviously $2^n \notin \mathrm{C}$ for $n = 1, 2, \ldots$.

Observe that $| \mathrm{C}_M |$ need not be much greater then $| \mathrm{C}_M^2 |$ as the following example shows: $\mathcal{A}_1 = \{2j - 1 : j \in \mathcal{N}\}$, $\mathcal{A}_2 = \{2j : j \in \mathcal{N}\}$. However the situation is different for $k \leq 3$ and for $k \geq 4$.

Theorem 2.

(i) *There is an absolute constant C so that if $k \leq 3$ then at any k-partition*

$$| \mathrm{C}_M | \geq \left[\frac{M}{2} \right] - 1 \quad \text{if} \quad M > C. \tag{17}$$

(ii) *If $k \geq 4$, there exists a k-partition such that*

$$| \mathrm{C}_M | < \frac{M}{2} - ck \log M, \tag{18}$$

where c is an absolute constant.

Proof of (i). *Case $k = 2$.*

Without loss of generality we can assume that $x \in \mathcal{A}_1$ for $1 \leq x \leq a$ and $a + 1 \in \mathcal{A}_2$.

Then $y \in \mathrm{C}$ for $3 \leq y \leq 2a - 1$. On the other hand for every $y > 0$ either $y + a \in \mathrm{C}$ or $y + a + 1 \in \mathrm{C}$.

Case $k = 3$

Suppose $2x - 1 \in \mathcal{A}_1$ if $1 \leq x \leq a$ and $2a + 1 \in \mathcal{A}_2$. Then

(19) $$2y \in \mathrm{C} \quad \text{if} \quad 2 \leq y \leq 2a.$$

We may assume that there is an $n > 2a$ such that

(20) $$2n \notin \mathrm{C} \quad \text{and} \quad 2n - 1 \notin \mathrm{C}$$

and

(21) $$| \mathrm{C}_{2n} | < \left[\frac{M}{2} \right]$$

Case 1 $2n \leq 6a$. Put $2n = 4a + 2t$, $(t \leq a)$. First we prove

$$2a + 2 \in \mathcal{A}_2.$$

Namely if $2a + 2 \in \mathcal{A}_1$, then $2x - 1 + 2a + 2 \in \mathrm{C}$ for $1 \leq x \leq a$. Hence

$$| \mathrm{C}_{2n} | > 2a + a$$

which contradicts (21).

Now suppose $2a + 2 \in \mathcal{A}_3$. Then $2n - (2a + 2) = 2a + 2t - 2 \in \mathcal{A}_1 \cup \mathcal{A}_2$. In case $2a + 2t - 2 \in \mathcal{A}_1$

$$2x - 1 + 2a + 2t - 2 \in \mathrm{C} \quad \text{for} \quad 1 \leq x \leq a.$$

This implies

$$\mathrm{C}_{2n} \geq 2a + a$$

which contradicts again to (21).

In case $2a + 2t - 2 \in \mathcal{A}_2$

$$2n - 1 = (2a + 1) + (2a + 2t - 2) \in \mathrm{C}$$

would follow, which contradicts to (20).

Thus $2a + 2 \in \mathcal{A}_2$.

Consider now the integers in $[2a + 2,\ 2a + 2t]$. For every y, $0 \leq y \leq 2t$

$$2a + y \in \mathcal{A}_3 \quad \text{implies} \quad 2a + 2t - y \in \mathcal{A}_3.$$

Therefore at least t integers in $[2a + 2,\ 2a + 2t]$ belong to $\mathcal{A}_1 \cup \mathcal{A}_2$.

If there is an even $x \in \mathcal{A}_1 \cup [2a + 2,\ 2a + 2t]$, then

$$4a < x + 2v - 1 < 4a + 2t = n \quad \text{for} \quad 1 \leq v \leq a.$$

Hence

$$| \mathrm{C}_{2n} | > 2a + a$$

which contradicts (21).

If all the t even integers in $[2a + 2,\ 2a + 2t]$ belong to $\mathcal{A}_2$, then for $1 \leq u \leq t - 1$

$$(2a + 2 + 2u) + 2a + 1 \in \mathrm{C}$$

and

$$(2a+2+2u)+2a+2 \in \mathbf{C}$$

This would imply

$$|\, \mathbf{C}_{2n} \,| > 3a.$$

This finishes the case when $2n \le 6a$.

Case 2 $2n > 6a$.

Since $2n \notin \mathbf{C}$, at least $\frac{n-2}{2}$ even numbers below $2n$ are in $\mathcal{A}_1 \cup \mathcal{A}_2$. Thus at least $\frac{n-2}{2} - a$ even numbers below $2n-2a$ are in $\mathcal{A}_1 \cup \mathcal{A}_2$. Therefore at least $\frac{n-2}{4} - \frac{a}{2} > \frac{n-2}{12}$ are in $\mathcal{A}_1$ resp. in $\mathcal{A}_2$. Adding to these numbers $2a-1$ or $2a+1$ we gain $\frac{n-2}{12}$ odd numbers in $\mathbf{C}$. Hence by Theorem 1

$$\mathbf{C}_{2n} > n - 6n^{\frac{15}{16}} + \frac{n-2}{12} > n \quad \text{if}$$

n is large enough.

Proof of (ii). We may suppose that $k = 4\ell$ where ℓ is odd. Define t_0 by

$$2^{t_0-1} \le 2\ell < 2^{t_0}$$

For $i = 1,2,\ldots,\ell$ we are going to define subsets $\mathcal{A}_{4i-j}$, $j = 0,1,2,3$ recursively. Let for $j = 1,3$

$$\mathcal{A}_{4i-j} \cap [1,2^{t_0}] = \{n : n \equiv i \pmod{\ell}, n \equiv \left[\frac{j}{2}\right] \pmod{2}\} \cap [1,2^{t_0}]$$

and

$$\mathcal{A}_{4i-j} \cap [1,2^{t_0}] = \emptyset \quad \text{if} \quad j = 0,2.$$

Assume now that $\mathcal{A}_{4i-j} \cap [1,2^t]$ have been defined for $j = 0,1,2,3$, $i = 1,\ldots,2\ell+1$. Let $r_i(t)$ defined by

$$2i \equiv 2^{t+1} + r_i(t) \pmod{2\ell}, \quad 0 \le r_i(t) < 2\ell.$$

Now we define $\mathcal{A}_{4i-j} \cap [2^t+1, 2^{t+1}]$ in the following way: let $2^t < n \le 2^{t+1}$. For $2^t < n < 2^{t+1}$ $n \in \mathcal{A}_{4i-3}$ iff n is even and

$$n \equiv i \pmod{\ell}, \quad 2^{t+1} + r_i(t) - n \notin \mathcal{A}_{4i-3} \cap [1,2^t], 2 \mid n$$

$n \in \mathcal{A}_{4i-2}$ iff n is even and

$$n \equiv i \pmod{\ell}, \quad n \notin \mathcal{A}_{4i-3},$$

$n \in \mathcal{A}_{4i-1}$ iff n is odd and

$$n \equiv i \pmod{\ell},\ 2^{t+1} + r_i(t) - n \notin \mathcal{A}_{4i-1} \cap [1,2^t], 2 \mid n$$

$n \in \mathcal{A}_{4i}$ iff n is odd and

$$n \equiv i \pmod{\ell},\ n \notin \mathcal{A}_{4i-1},$$

Then clearly the sets $\mathcal{A}_{4i-j}$, $1 \le i \le \ell$, $0 \le j \le 3$ give a 4ℓ-partition of $\mathcal{N}$. Furthermore it can be seen easily that all the monochromatic sums

$a_1 + a_2$, $a_1 \neq a_2$ are even and none of these sums is equal to a number of the form $2^t + 2^j$ where $t > t_0$ and $0 \le j \le \ell - 1$. This completes the proof of Theorem 2.

By Theorem 1, there are more than $\frac{M}{2} - c_1 M^{1-2^{k-1}}$ integers in $[1, M]$ which have a monochromatic representation in the form (2), and by Theorem 2, the number of these integers can be less than $\frac{M}{2} - c_2 k \log M$. It follows from a result of Erdős and Sárközy (Theorem 8 in [5]) that if $k \in \mathcal{N}$, $M \in \mathcal{N}$, $M > M_0(k)$, $t \in \mathcal{N}$ and $M^{2/3}(\log M)^2 < t \le M$, then almost all the sets $\mathcal{B}$ with $\mathcal{B} \subset [1, M]$, $|\mathcal{B}| = t$ are such that for every k-partition of $[1, M]$ there is (at least one) element in $\mathcal{B}$ which has a monochromatic representation in the form (2). (In fact, the following sharper statement is true: almost all of these sets $\mathcal{B}$ are such that for every $\mathcal{A}$ with $\mathcal{A} \subset [1, \frac{M}{2}]$ and $|\mathcal{A}| > \frac{1}{k}[M/2]$, there is an element in $\mathcal{B}$ which can be represented in the form (2) with $a \in \mathcal{A}$, $a' \in \mathcal{A}$.) Ruzsa [16] proved that if $f(x) \to +\infty$, then there exists an infinite sequence $\mathcal{D}$ of positive integers such that $D(x) = \sum_{\substack{d \le x \\ d \in \mathcal{D}}} | = 0(f(x)(\log x)^2)$, and if $\mathcal{A}$ is a sequence of positive integers with positive upper asymptotic density, then $\mathcal{D}$ intersects the set of the integers of the form $a + a'$ where $a \in \mathcal{A}$, $a' \in \mathcal{A}$. These results suggest that the upper bound $\frac{M}{2} - ck \log M$ is closer to the truth than the lower bound.

Recently Balog, Fürstenberg, Sárközy, Stewart, Lagarias, Odlyzko, Schearer [1], [7], [13], [14], [17], [18], [19] and others have studied the solvability of the equations

$$
\begin{aligned}
a - a' &= x^2 \\
a - a' &= p - 1 \\
a + a' &= x^2 \\
a + a' &= px, \quad x \text{ "small" } (= 0(1))
\end{aligned}
$$

with $a, a' \in \mathcal{A}$ where $\mathcal{A}$ is a "dense" sequence of positive integers. These results and Hindman's theorem [2], [11] led us to consider the corresponding "monochromatic" questions.

Theorem 1 implies that e.g. the equations

$$
\begin{aligned}
a_1 + a_2 &= 2p \\
a_1 + a_2 &= p - 1
\end{aligned}
$$

have monochromatic solutions with $a_1 \neq a_2$.

Our result is not strong enough to obtain for arbitrary k that

$$a_1 + a_2 = x^2$$

has a monochromatic solution with $a_1 \neq a_2$. However a simple argument leads to

Theorem 3. *If $k \leq 3$, then for any k-partition of $\mathcal{N}$ there are infinitely many squares in* $\mathbf{C}$.

Proof. We use the following simple (and well known)

Lemma 2. *For every $\varepsilon > 0$ there are infinitely many integers n so that*

$$n = x^2 + y^2$$

has at least three (in fact arbitrary many) integer solutions where

$$x^2, y^2 \in \left[\frac{n}{2}(1-\varepsilon),\ \frac{n}{2}(1+\varepsilon)\right].$$

Now let

$$x_1^2 + x_6^2 = x_2^2 + x_5^2 = x_3^2 + x_4^2$$

with $x_i \in \left[\frac{n}{2}(1-\varepsilon),\ \frac{n}{2}(1+\varepsilon)\right],\ \ 1 \leq i \leq 6.$

Then an easy calculation shows, that the system

$$\begin{aligned} u_1 + u_2 &= x_1^2 \\ u_3 + u_4 &= x_6^2 \\ u_2 + u_3 &= x_2^2 \\ u_1 + u_4 &= x_5^2 \\ u_1 + u_3 &= x_3^2 \\ u_2 + u_4 &= x_4^2 \end{aligned}$$

in $u_i (1 \leq i \leq 4)$ has a solution in distinct positive numbers. Since at least two of the u_i's belong to the same class, one of the x_i^2 $(1 \leq i \leq 6)$ squares must have a monochromatic representation.

If we have some information on the structure of the classes $\mathcal{A}_i$ in the given partition then the lower bound given for the integers that have a monochromatic representation in form (2) can be sharpened. In fact we have

Theorem 4.

(i) *For every $\varepsilon > 0$ and k there exists an $M_0(\varepsilon, k)$ such that if we have a k-partition of $\mathcal{N}$ where every class contains both even and odd integers then*

$$|\, \mathbf{C}_M \,| > \left(\frac{1}{2} + \frac{1}{2k} - \varepsilon\right) M \text{ if } M > M_0(\varepsilon, k).$$

(ii) *For every $k \in \mathcal{N}$ there is a k-partition of $\mathcal{N}$ so that every class contains both even and odd integers and*

$$|\, \mathbf{C}_M \,| < \left(\frac{1}{2} + \frac{1}{k}\right) M + 1.$$

Proof.

(i) can be proved by the method used in the proof of Theorem 2,

(ii) follows from the following construction: for $i = 1, 2, \dots, k$ let

$$\mathcal{A}_i = \{n : n \equiv 2i \pmod{2k}\} \cup \{n : n \equiv 1 - 2i \pmod{2k}\}.$$

It is easy to see that this k-partition of $\mathcal{N}$ has the desired properties.

The Case $f(x_1, x_2) = | rx_1 + sx_2 |$.

Let r, s be integers. As before, let **C** denote the set of integers which have a monochromatic representation in the form

(22) $$n = | ra_1 + sa_2 | \quad \text{with} \quad a_1 \neq a_2.$$

Let $\mathbf{C}_M =: C \cap [1, M]$. The following result is merely a simple modification of Theorem 1.

Theorem 5. *Let $r \neq 0$, $s \neq 0$, $r + s \neq 0$. Put $| r + s | = m$. For every $\varepsilon > 0$, k, r, s and for every k-partition*

$$| \mathbf{C}_M | \geq (1 - \varepsilon)\frac{M}{m}.$$

This can not be essentially improved, since choosing

(23) $$k = m \text{ and } \mathcal{A}_i = \{n : n \equiv i \pmod{m}\},\ 1 \leq i \leq m$$

only the multiples of m have a monochromatic representation in the form (22)

Note furthermore that Theorem 5 does not cover the case of the *differences* $a_1 - a_2$. Namely, in this case the density of the integers having a monochromatic representation in the form (22) need not be greater than a positive absolute constant. To see this, let us consider a large integer m and define the partition as in (23). Then only the multiples of m have a monochromatic representation in the form (22) so that their density is $\frac{1}{m}$ which $\to 0$ if $m \to \infty$.

Proof. Assume that there are more than $\varepsilon\frac{M}{m}$ positive multiples of m in $[1, M]$ which do not have a monochromatic representation in the form [22]. Then by Szemerédi's theorem [20], for $M > M_0(k, \varepsilon, r, s)$ their set must contain an arithmetic progression of $2(| r | + | s |)k + 1$ terms; let us write this arithmetic progression (all whose terms are multiples of m) in the form

(24) $$um - (| r | + | s |)kv,\ um - ((| r | + | s |)k - 1)v, \dots,\ um + (| r | + | s |)kv.$$

Let us consider the integers $u, u+v, \ldots, u+kv$. By the pigeon hole principle, two of them, say $a_1 = u+iv$ and $a_2 = u+jv$ (where $i \neq j$) belong to the same class. Then

$$ra_1 + sa_2 = r(u+iv) + s(u+jv) = (r+s)u + (ri+sj)v.$$

Here we have

$$| ri + sj | \leq | r | k + | s | k = (| r | + | s |)k.$$

Since $| r+s | = m$ and all the numbers in (24) are positive, $| ra_1 + sa_2 |$ is equal to one of the numbers in (24). But this contradicts the fact that none of these numbers has a monochromatic representation in the form (22), and the proof is completed.

2. Some Unsolved Problems

Problem 1. Do there exist α and β which depend only on k, so that for an arbitrary k-partition

$$| \mathbf{C}_M | > \frac{M}{2} - (\log M)^{\alpha(k)}$$

or even more $| C_M^2 | > \frac{M}{2} - (\log M)^{\beta(k)}$.

Problem 2. Let $f(x)$ be a polynomial of integer coefficients such that 2 is a prime divisor of it. Is it true that for any k-partition for some x (or for infinitely many x)

$$a_1 + a_2 = f(x),$$

have a monochromatic solution with $a_1 \neq a_2$?

Problem 3. Is it true that for every k-partition of $[1, M]$ almost all the even integers $2n$ in $[1, M]$ have more than $c(k)n$ monochromatic representations in form (2)? (Perhaps this holds with $c(k) = \frac{c_1}{k}$.)

Problem 4 a) For a given k-partition let $n_1 < n_1 < \ldots$ be the sequence of those integers which have a monochromatic representation in form (2). ($\mathbf{C} = \{n_i\}$). What can be said about the structure of the sequence $\{n_i\}$? (For example it is easy to see that $| n_{i+1} - n_i | < 2k$.)

b) The complementary problem is to study the structure of the set $\mathcal{B} = \mathcal{N} - \mathbf{C}$ (the set of those integers which do not have a monochromatic representation in form (2)).

Let $\mathcal{G}(\mathcal{N}; E)$ be the graph with edgeset $\{(x, y) \mid x + y \in \mathcal{B},\ x, y \in \mathcal{N}\}$. Obviously at any k-partition the chromatic number of $\mathcal{G}(\mathcal{N}; E)$ is $\leq k$. Basically this was used in the proofs above.

Problem 5. So far we have studied monochromatic representations in form (1) in the special case when $f(x_1, \dots, x_t)$ is a linear polynomial and $t = 2$. In the paper Erdős–Sárközy [6] the case $f(x_1, x_2) = x_1 x_2$ is considered.

What can one say on general polynomials $f(x_1, \dots, x_t)$ (whose coefficients are integers)? What can be said in the most important special case when $f(x_1, x_2, \dots, x_t)$ is of the form $g(x_1) + \dots + g(x_t)$?

As Ruzsa [15] observed, if

$$f(x_1, x_2, x_3, x_4) = x_1^2 + x_2^2 + x_3^2 + x_4^2$$

then for every k-partition

$$| C_M | > c(k) \cdot M$$

and $| C_M | > cM$ cannot hold with an absolute constant c.

Acknowledgement. We would like to thank to I. Ruzsa for his helpful comments.

References

[1] A. Balog and A. Sárközy, On sums of sequences of integers, *II, Acta Math. Hung.* **44** (1984), 169-179.

[2] J. Baumgartner, A short proof of Hindman's theorem, *J.Comb. Th.* Ser. A **17** (1974), 384-386.

[3] D. Berend, Joint ergodicity and mixing, *J. Analyse Math.* **45** (1985), 255-284.

[4] P. Erdős, Some unsolved problems, *MTA MKI Közl.* **6** (1961), 221-254.

[5] P. Erdős and Sárközy, On differences and sums of integers, I, *J. Number Theory* **10** (1978), 430-450.

[6] P. Erdős and A. Sárközy, On monochromatic products, *to appear.*

[7] H. Fürstenberg, Ergodic behavior of diagonal measures and a theorem of Szemerédi on arithmetic progressions, *J. Analyse Math.* **31** (1977), 204-256.

[8] H. Fürstenberg and B. Weiss, Topological dynamics and combinatorial number theory, *J. Analyse Math.* **34** (1978), 61-85.

[9] R.K. Guy, *Unsolved Problems in Number Theory,* Springer-Verlag, 1981.

[10] D. Hilbert, Über die Irreduzibilität ganzer rationaler Functionen mit ganzzahligen Koefficienten, *J. Reine Angew. Math.* **110** (1892), 104-129.

[11] N. Hindman, Finite sums from sequences within cells of a partition of N, *J. Comb. Th.* Ser. A **17** (1974), I-II.

[12] J.C. Lagarias, A.M. Odlyzko and J.B. Shearer, On the Density of Sequences of Integers the Sum of No Two of which is a Square, I, Arithmetic Progressions, *J. Comb. Th.* Ser. A **33** (1982), 167-185.

[13] J.C. Lagarias, A.M. Odlyzko and J.B. Shearer, On the Density of Sequences of Integers the Sum of No Two of which is a Square, II, General Sequences, *J. Comb. Th.* Ser. A **34** (1983), 123-139.

[14] C. Pomerance, A. Sárközy and C.L. Stewart, On divisors of sums of integers, II, *Pacific J. Math., to appear.*

[15] I.Z. Ruzsa, *oral communication.*

[16] I.Z. Ruzsa, Probabilistic constructions in additive number theory, *Soc. Math. France Astérique* **147-148** (1987), 173-182.

[17] A. Sárközy, On difference sets of sequences of integers, I, *Acta Math. Hung.* **31** (1978), 125-149.

[18] A. Sárközy, On difference sets of sequences of integers, III, *Acta Math. Hung.* **31** (1978), 355-386.

[19] A. Sárközy and C.L. Stewart, On divisors of sums of integers, II, *Math. Annalen.*

[20] E. Szemerédi, On sets of integers containing no k elements in arithmetic progression, *Acta Arithm.* **27** (1975), 199-245.

P. Erdős
Math. Inst. of the
Hung. Acad. of Sci.
Budapest

A. Sárközy
Math. Inst. of the
Hung. Acad. of Sci.
Budapest

V.T. Sós
Math. Inst. of the
Hung. Acad. of Sci.
Budapest

5. Discrepancy of Sequences in Discrete Spaces

Ph. Flajolet, P. Kirschenhofer and R.F. Tichy

Abstract

In this article some combinatorial properties of 0.1–strings are established. A special sequence with small discrepancy (with respect to increasing block length) is constructed by means of De Bruijn graph. Applying a general result of W. Philipp a central limit theorem is derived for the local discrepancy.

1. Introduction

The study of "uniformly distributed" sequences in discrete spaces is motivated by several applications such as the construction of random sequences (compare D.E. Knuth [5, Section 3.5]) and some problems in linguistics (compare E. Hlawka [2]). In the paper [4] two of the authors have introduced the following notion of uniform distribution (for short: u.d.) which was stimulated by definition Q 1 of [5, Section 3.5]:

Let $A = \{a_1, \ldots, a_n\}$ be a finite alphabet ($\alpha \geq 2$) and $s(N)$ a (weakly) increasing sequence of positive integers with $s(N) \leq N$. Then a sequence $\omega = (x_n)_{n=1}^{\infty}$ with elements in A is called $[s(N)]$-u.d. iff

$$D^{[s(N)]}(\omega_N) := \max_{u \in A^{s(N)}} \left| D^{[s(N)]}(u, \omega) \right| = o(\alpha^{-s(N)}) \quad (N \to \infty) \text{ with}$$

(1.1)

$$D^{[s(N)]}(u, \omega) := \left| \frac{[\omega_N; u]}{N - s(N) + 1} - \alpha^{-s(N)} \right| \quad (\text{for } u \in A^{s(N)}),$$

where $\omega_N = x_1 \ldots x_N$ is the initial string of length N of ω and $[\omega; u]$ denotes the number of occurrences of u as a (closed) subblock in the word ω (for example $[00101; 01] = 2$). $D^{[s(N)]}(\omega_N)$ is called the discrepancy of ω and $D^{[s(N)]}(u; \omega)$ the local discrepancy.

Our aim in this article is twofold: on the one hand we generalize the results of [4] from the case $\alpha = 2$ to general α and on the other hand we give a much sharper version of a metric theorem concerning $[s(N)]$-u.d. sequences. In section 2 we present the results concerning $[s(N)]$-uniform distribution. For further references compare the survey article [3].

2. Results

It is an easy observation that $s(N) \le \log_\alpha N$ (for sufficiently large N) is a necessary condition for the existence of an $[s(N)]$-u.d. sequence ω in A. Furthermore we establish the following result.

Proposition 2.1. *Let $s(N) \le t(N)$ with $t(N)\alpha^{t(N)} = o(N)$ and let ω be $[t(N)]$-u.d. Then ω is also $[s(N)]$-u.d.*

Proof. An elementary combinatorial argument shows

$$[\omega_n; u_1 \dots u_{s(N)}] = \sum_{v \in A^{t(N)-s(N)}} [\omega_{N+t(N)-s(N)}; u_1 \dots u_{s(N)} v].$$

Hence

$$\left| \frac{[\omega_N; u_1 \dots u_{s(N)}]}{N - t(N) + 1} - \alpha^{-s(N)} \right| \le \alpha^{t(N)-s(N)} D^{[t(N)]}(\omega_N),$$

and we obtain

$$D^{[s(N)]}(\omega_N) \le \alpha^{t(N)-s(N)} \left(D^{[t(N)]}(\omega_N) + \frac{t(N) - s(N)}{N - t(N) + 1} \right), \tag{2.2.}$$

from which the result follows immediately. □

The construction of $[s(N)]$-u.d. sequences is non-trivial if $\lim_{N\to\infty} s(N) = \infty$. One source for such a construction may be the theory or normal numbers (cf. [5,Sec,3.5.]). In the following we present a different construction based on the idea of De Bruijn graph $X_s (s \ge 2)$:

Let the vertex set $V(X_s)$ be A^{s-1} and the set $E(X_s)$ of directed edges be given by A^s in the following way: the edge $x_1 \dots x_s$ is the directed connection of the vertex $x_1 \dots x_{s-1}$ with the vertex $x_2 \dots x_s$.

Thus each vertex has indegree and outdegree both equal to α. Hence X_s is Eulerian and using all different closed Eulerian lines of X_s we obtain all different sequences on A (of length $\alpha^s + s - 1$) that contain each word of A^s exactly once:

Considering a special Eulerian line of X_s with the edges $e_1 \dots e_t (t = \alpha^s)$, $e_i = x_{i,1} \dots x_{i,s} \in A^s$,we define the string $x_{1,1} \dots x_{1,s} x_{2,s} x_{3,s} x_{4,s} \dots x_{t,s}$. Let $y_{1,s} \dots y_{\alpha^s+s-1,s}$ be such a string starting with a_1^s. Then we set

$$\omega = a_1 \dots a_\alpha y_{1,2} \dots y_{\alpha^2+1,2} y_{2,3} \dots y_{\alpha^3+2,3} \dots y_{s-1,s} \dots y_{\alpha^s+s-1,s} \dots \tag{2.3.}$$

Theorem 2.4. *Let ω be a sequence of type (2.3) and $s(N) = o(\log\log N)$. Then ω is $[s(N)]$-u.d.*

Proof. Since the arguments are similar to the proof of the special case $\alpha = 2$ in [4, Theorem 2.9.], we present only a sketch. W.1.o.g. we assume $\lim_{N\to\infty} s(N) =$. Let N be such that $\frac{\alpha^m-\alpha}{\alpha-1} < N \le \frac{\alpha^{m+1}-\alpha}{\alpha-1}$. Then, by the construction of ω, we have for any $u \in A^s$ $(s = s(N))$

$$\frac{\alpha^{m-s}-1}{\alpha-1} + [\tilde\omega_N; u] \le [\omega_m; u] \le \frac{\alpha^s + \alpha^{m-s}}{\alpha-1} + [\tilde\omega_N; u],$$

where $\tilde\omega_N = x_{\frac{\alpha^m-1}{\alpha-1}} \dots x_Nfor\omega = x_1x_2x_3\dots$ Thus it is sufficient to show that

$$\frac{\frac{\alpha^m}{\alpha-1} + \alpha^s[\tilde\omega_N; u]}{N-s+1} - 1 = o(1) \tag{2.5.}$$

uniformly in $u \in A^{s(N)}$. By the above construction, ω_N is the initial part of a string containing each m-element string exactly once. If $e_1, e_2, \dots e_{\alpha^m}$ denote the edges of the Eulerian line of X_m we have

$$\frac{\sum[e_i; u]}{m-s+1} \le [\tilde\omega_N; u] \le \frac{\sum[e_i; u]}{m-s+1} + O(m),$$

where the sums are extended over all i with $1 \le i \le N - \frac{\alpha^m-\alpha}{\alpha-1} - s + 1$. For fixed u and $\varepsilon > 0$ we distinguish two cases of edges e_i

$$\left|\frac{[e_i; u]}{m-s+1} - \alpha^{-s}\right| \le \varepsilon, \tag{i}$$

$$\left|\frac{[e_i; u]}{m-s+1} - \alpha^{-s}\right| > \varepsilon. \tag{ii}$$

Considering the random variables $\xi_i = \delta_{x_i \dots x_{i+s-1}, u}$ (where $\omega = x_1x_2\dots$ and δ the Kronecker symbol) ξ_i and ξ_j are independent for $|i-j| \ge s$. By a slight modification of the well-known Bernoulli-Chebyshev inequality for independent random variables we obtain that the number L of strings e_i falling into case (ii) fulfills

$$L = O\left(\frac{s\alpha^m}{\varepsilon^2(m-s+1)}\right). \tag{2.6.}$$

Furthermore we have

(2.7.)

$$(N - \frac{\alpha^m-\alpha}{\alpha-1} - s + 1 - L)(\alpha^{-s} - \varepsilon) \le [\tilde\omega_N; u] \le (N - \frac{\alpha^m-\alpha}{\alpha-1})(\alpha^{-s}+\varepsilon) + L + O(m).$$

Choosing $\varepsilon = \varepsilon(N) = \alpha^{-2s(N)}$ and combining (2.6) and (2.7) yields (2.5). Thus the proof of our theorem is completed. □

Remark 2.8. In a forthcoming paper [1] M. Goldstern gives a refinement of the above construction. Instead of (2.3) a sequence ω^* is defined by running through the Eulerian line originating from X_k exactly α^k times. In the case $\alpha = 2$ it is shown that ω^* is $[s(N)]$-u.d. for $s(N) \leq c \log_2 N$ with $c < \frac{1}{2}$. This construction can be refined: a sequence ω^{**} can be found which is $[s(N)]$-u.d. for $s(N) \leq c \log_2 N$ with $c < 1$ (cf.[1]).

In the following we establish a quantified version of our Theorem 2.5. of [4]; we establish a central limit theorem (prob denotes the probability in the sense of product measure on A^∞, where the elements of A are equally likely).

Theorem 2.9. *Let $s(N)$ be a weakly increasing and unbounded sequence of positive integers with $\lim_{N\to\infty} \frac{s(N)^2 \alpha^{s(N)}}{N} = 0$. Then for every $u \in A^{s(N)}$*

$$\text{prob}\left\{ \frac{\sqrt{N}}{\sigma_N} D^{[s(N)]}(u;\omega) < y \right\} =$$

$$= \frac{1}{\sqrt{2\pi}} \int_{-\infty}^{y} e^{-t^2/2} dt + O\left(\frac{s(N)\alpha^{s(N)/2}}{\sqrt{N}} \right) + O\left(\sqrt{\frac{\log s(N)}{s(N)}} \right)$$

with O-constants only depending on $\alpha; \sigma_N^2 = O(\alpha^{-s(N)})$ denotes the variance of $D^{[s(N)]}(u;.)$.

Proof. The proof is essentially based on the ideas of the proof of Satz 1 in [6] by W. Philipp and because of its length we divide it into several lemmata. Let $X_{n,N} = X_{n,N,u} = \delta_{x_n \ldots x_{n+s(N)-1},u} - \alpha^{-s(N)}$ with $\omega = (x_0, x_1, \ldots)$ and $u \in A^{s(N)}$, $0 \leq n \leq N$, so that $\sum_{0 \leq n \leq N} X_{n,N} = [x_0, \ldots, x_{N+s(N)-1}; u] - (N+1)\alpha^{-s(N)} = [x_0, \ldots, x_N; u] - (N - s(N) + 1)\alpha^{-s(N)} + O(s(N))$. Then for fixed N and u the $X_{n,N}$ are equally distributed random variables with expectation $E(X_{n,N}) = O$.

Lemma 1. *For all integers $r \geq 2$, $0 \leq i_1 < \ldots < i_j \leq N$, $1 \leq j < r$, $p_\nu \geq 0$ $(1 \leq \nu \leq r)$ we have*

$$|E(X_{i_1,N}^{p_1} \cdots X_{i_r,N}^{p_r}) - E(X_{i_1,N}^{p_1} \cdots X_{i_j,N}^{p_j}) E(X_{i_{j+1},N}^{p_{j+1}} \cdots X_{i_r,N}^{p_r}) \leq$$
$$\leq \alpha^{-s(N)-(i_{j+1}-i_j)} (1 - \alpha^{-s(N)})^{\Sigma p_\nu}.$$

Proof. For short we write $X_{i,N} = X_i = \delta_i - \alpha^{-s(N)}$ and obtain by the Binomial Theorem $X_{i,N}^p \gamma(p) + \delta_i \rho(p)$, where

$$\gamma(p) = (-1)^p \alpha^{-sp},\ \rho(p) = (1-\alpha^{-s})^p - \gamma(p)\ (s = (s(N)).$$

Then with

$$M = \{i_1, \ldots, i_r\},\ K_1 = \{i_1, \ldots, i_j\} \cap K,\ K_2 = \{i_{j+1}, \ldots, i_r\} \cap K$$

(for any $K \subseteq M$)

$$E(X_{i_1}^{p_1} \dots X_{i_r}^{p_r}) - E(X_{i_1}^{p_1} \dots X_{i_j}^{p_j}) E(X_{i_{j+1}}^{p_{j+1}} \dots X_{i_r}^{p_r}) =$$
$$= \sum_{K \subseteq M} \prod_{i \in M-K} \gamma(p_i) \prod_{i \in K} \rho(p_i) [E(\prod_{i \in K} \delta_i) - E(\prod_{i \in K_1} \delta_i) E(\prod_{i \in K_2} \delta_i)].$$

Let

$$i_m = \max\{i_k : i_k \in K_1\}, \; i_M = \min\{i_k : i_k \in K_2\} \text{ and } K_1, K_2 \neq \emptyset.$$

Case 1. $i_M - i_m \geq s$. The above difference of expected values (in edged brackets) is 0, since the concerned random variables are independent.
Case 2. $i_M - i_m < s$. Observing that $\prod_{j \in K} \delta_i = 1$ fixes the elements $x_{i_m} \dots x_{i_M + s - 1}$ of the sequence ω, we have

$$E(\prod_{i \in K} \delta_i) \leq \alpha^{-s-(i_n - i_m)} \leq \alpha^{-s-(i_{j+1} - i_j)}.$$

By a similar argument we obtain

$$E(\prod_{i \in K_1} \delta_i) E(\prod_{i \in K_2} \delta_i) \leq \alpha^{-2s} \leq \alpha^{-s-(i_{j+1} - i_j)}.$$

If $K_1 = \emptyset$ or $K_2 = \emptyset$, then the above difference (in the edged brackets) is obviously 0. Thus we have

$$|E(X_{i_1}^{p_1} \dots X_{i_r}^{p_r}) - E(X_{i_1}^{p_1} \dots X_{i_j}^{p_j}) E(X_{i_{j+1}}^{p_{j+1}} \dots X_{i_r}^{p_r})| \leq$$
$$\leq \alpha^{-s-(i_{j+1} - i_j)} \sum_{k \subseteq M} \prod_{i \in M-K} \gamma(p_j) \prod_{i \in K} \rho(p_j) =$$
$$= \alpha^{-s-(i_{j+1} - i_j)} \prod_{k=1}^{r} (\gamma(p_k) + \rho(p_k)) = \alpha^{-s-(i_{j+1} - i_j)} (1 - \alpha^{-s})^{\Sigma p_k}$$

and the proof of Lemma 1 is complete. □

As an immediate consequence we have

Lemma 2. *Under the assumptions of Lemma 1:*

$$|E(X_{i_1,N}^{p_1} \dots X_{i_r,N}^{p_r}) - E(X_{i_1,N}^{p_1} \dots X_{i_j,N}^{p_j}) \cdot E(X_{i_{j+1},N}^{p_{j+1}} \dots X_{i_r,N}^{p_r})|$$
$$\leq \alpha^{-(i_{j+1} - i_j)} E(|X_{O,N}|^{\sum p_\nu}).$$

Hence the essential condition in (1.2) of [6, Satz 1]. is fulfilled with $L(x) = 1, c(t) = t^{-\psi(t)}$, $\psi(t) = \frac{1}{\log t} \log \alpha$. Instead of Hilfssatz 1.1. in [6] we prove

Lemma 3.

(i) $E(X_{O,N}^2) = \alpha^{-s(N)} - \alpha^{-2s(N)}$,

(ii) $\tilde{\delta}_N^2 := \frac{1}{N}(\sum_{n\le N} X_{n,N,u}^2) = \alpha^{-s(N)} + 2\alpha^{-s(N)} \sum_* \alpha^{-t} + O(s(N)\alpha^{-2s(N)})$ *where the sum* $(*)$ *is taken over all* t *with* $1 \le t \le s(N) - 1$ *such that* $u_0 \dots u_{s(N)-t-1} = u_t \dots u_{s(N)-1} (u = u_0 u_1 u_2 \dots)$. *(The O-constant may only depend on* α.*)*

Proof. (i) is trivial by the definition of the random variables. In order to prove (ii) we observe

$$\tilde{\sigma}_N^2 = E(X_{O,N}^2) + 2 \sum_{1\le t\le N} E(X_{O,N} X_{t,N}) - \frac{1}{N} \sum_{1\le t\le N} tE(X_{O,N} X_{t,N}).$$

Now

$$E(X_{O,N} X_{t,N}) = \begin{cases} 0 & \text{for } t \le s(N), \\ -\alpha^{-2s(N)} + \alpha^{-s(N)-t} & \text{for } 1 \le t \le s(N) \\ & \text{-and condition } (*) \text{ holds}, \\ -\alpha^{-2s(N)} & \text{for } 1 \le t \le s(N) - 1 \\ & \text{and condition } (*) \text{ does not hold}, \end{cases}$$

from which (ii) follows by an easy calculation. □

As in [6] we define the new random variables $Y_{i,N}^* (= Y_{i,n,u}^*)$ and $V_{i,N}^* (= V_{i,n,u}^*)$ in the following way:
Let $h = \lfloor N^{\frac{1}{2}} \rfloor$ (greatest integer $\le N^{\frac{1}{2}}$) and $k = \lfloor N^{\frac{1}{4}} \rfloor$. Then

$Y_{1,N}^* = X_{O,N} + \dots + X_{h-1,N}$ $\qquad$ $V_{1,N}^* = X_{h,N} + \dots + X_{h+k-1,N}$

...... $\qquad$ ......

$Y_{l,N}^* = X_{(l-1)(h+k),N} + \dots X_{lh+(l-1)k-1,N}$ $\qquad$ $V_{l,N}^* = X_{lh+(l-1)k,N} + \dots + X_{N,N}$

so that $lh + (l-1)k \le N < (h+k)l$; thus $\frac{1}{2}N^{\frac{1}{2}} \le l \le N^{\frac{1}{2}}$ for sufficiently large N. We further set

$$S_N = Y_N + V_N := \sum_{i\le l} Y_{i,N}^* + \sum_{i\le l} V_{i,N}^* = \sum_{0\le n\le N} X_{n,N}.$$

Lemma 4.

(i) $\frac{1}{N} E(Y_N^2) = \tilde{\sigma}_N^2 + O(N^{-1/y}\alpha^{-s(N)})$,

(ii) $E(Y_{1,N}^{*2}) = h\tilde{\sigma}_N^2 + O(\alpha^{-s(N)})$.

Proof. We have

$$E(Y_N^2) = lE(\tilde{Y}_{1,N}^{*2}) + \sum_{i\ne j} E(Y_{i,N}^* Y_{j,N}^*),$$

and

$$E(Y_{1,N}^{*2}) = h(E(X_{0,N}^2) + 2 \sum_{1 \le t \le s(N)-1} E(X_0 X_t) - \frac{1}{h} \sum_{1 \le t \le s(N)-1} tE(X_0 X_t)) =$$
$$= 4\tilde{\sigma}_N^2 + O(\alpha^{-s(N)}).$$

By $E(Y_{i,N}^* Y_{j,N}^*) = O(h^2 \alpha^{-s-k})$ we obtain $\frac{1}{N} E(Y_N^2) = \frac{lh}{N} \tilde{\sigma}_N^2 + O(\frac{l}{N} \alpha^{-s(N)}) + O(\frac{l^2 h^2}{N} \alpha^{-s(N)-k})$. Since $\frac{lh}{N} = 1 + O(N^{-\frac{1}{4}})$ the assertions of the lemma follow immediately. □

Lemma 5. *Let p be an integer such that $1 \le p \le \frac{1}{32} \frac{s(N)}{\log_\alpha s(N) + (\log\log\alpha / \log\alpha)} \le \frac{1}{32} \frac{\log N}{\log\log N}$ with $\psi(N^{\frac{1}{16}}) \ge 4(p+1)$. Then*

(i) $E(Y_N^{2p}) = \frac{(2p)!}{2^p p!} N^p \tilde{\sigma}_N^{2p} + O(C^{2p} E_N^{2p} N^{p-3/8+c_1})$,

(ii) $E(Y_N^{2p+1}) = O(C^{2p+1} E_N^{2p+1} N^{p+3/8+c_2})$,

(iii) $E(Y_N^{2p}) = O(C^{2p} E_N^{2p} N^{(3p/4)+1/32})$,

where $C > 1$ is an absolute constant, $c_1, c_2, \ldots$ are arbitrary small positive constants and $E_N = E(X_{0,N}^{2p})^{1/2p}$. (Note $\psi(t) = \frac{t}{\log t} \log \alpha$).

Proof. To prove (i) we expand Y_N^{2p} by the multinomial theorem and obtain

$$\int (\sum_{i \le l} Y_{i,N}^*)^{2p} = (\sum{}' + \sum{}'' + \sum{}''') \frac{(2p)!}{p_1! \ldots p_l!} E(Y_{1,N}^{*p_1} \ldots Y_{l,N}^{*p_l}),$$

where $p_i \ge 0$ with $\sum p_i = 2p$; $\sum'$ contains all terms with $p_i = 0$ or 2 $(1 \le i \le l)$, $\sum''$ all terms with $p_i \ne 1 (1 \le i \le l)$ such that there exists an index j with $p_j \ne 0$ or $p_j \ne 2$; $\sum'''$ contains the remaining terms. Then

$$\sum{}' = \frac{(2p)!}{2^p} \sum_{1 \le i_1 < \ldots i_p \le l} E(Y_{i_1,N}^{*2} \ldots Y_{i_p,N}^{*2}).$$

Now,

$$|E(Y_{1,N}^{*p_1} \ldots Y_{t,N}^{*p_t}) - \prod_{i \le t} E(Y_{i,N}^{*p_i})| \le (hE_N)^{2p} \alpha^{-k} t.$$

For integral $p, p_i \ge 1$ with $\sum p_i = 2p$; the proof of this inequality runs verbally along the same lines as the proof of Hilfssatz 1.5. in [6]. By the last inequality

$$\sum{}' = \frac{(2p)!}{2^p} \binom{l}{p} (E(Y_{1,N}^{*2})^p + O((hE_N)^{2p} p \alpha^{-k}).$$

From $|\binom{l}{p} - \frac{l^p}{p!}| \le \frac{l^{p-1}}{(p-2)!}$ and Lemma 4 we derive

$$\sum{}' - \frac{(2p)!}{2^p p!} N^p \tilde{\sigma}_N^{2p} = O(C^{2p} E_N^{2p} N^{p-1/2+c_3}),$$

since $E_N^{2p} = \alpha^{-s(N)} + O(p\alpha^{-2s(N)}) + O(\alpha^{-2s(N)p})$, so that $\tilde{\sigma}_N^2 = O(E_N^{2p})$. The estimates for $\sum''$ and $\sum'''$ can be taken immediately from the proof of

Hilfssatz 1.6. in [6]:

$$\sum'' = O(C^{2p} E_N^{2p} N^{p-3/8+c_4}),$$

$$\sum''' = O(C^{2p} E_N^{2p} N^{p-1}).$$

By addition of the estimates we gain (i). (ii) and (iii) can be proved in a similar way. □

An immediate consequence is the following

Lemma 6. *Under the assumptions of Lemma 5 we have*

(i) $E(S_N^{2p}) = \frac{(2p)!}{2^p p!} N^p \tilde{\sigma}_N^{2p} + O(N^{p-3/32} C^{2p} E_N^{2p})$,

(ii) $E(S_N^{2p+1}) = O(N^{p+13/32} C^{2p+1} E_N^{2p+1})$.

In order to complete the proof of Theorem 2.9. we calculate the characteristic function of $\frac{S_N}{\tilde{\sigma}_N \sqrt{N}}$ and get for all integers Q with $1 \le Q \le \frac{1}{32} \frac{s(N)}{\log_\alpha s(N) + (\log\log\alpha/\log\alpha)}$, $\psi(N^{1/16}) \ge 4(Q+1)$:

$$\phi_N(t) := \int \exp(\frac{S_N}{\tilde{\sigma}_N \sqrt{N}} it) =$$

$$= \sum_{p=0}^{Q-1} \frac{((-1)^p t^{2p}}{(2p)!} E(\frac{S_N^{2p}}{\tilde{\sigma}_N^{2p} N^p}) + O\left(\frac{t^{2Q}}{(2Q)!} \frac{E(S_N^{2Q}}{\tilde{\sigma}_N^{2Q} N^Q}\right) +$$

$$+ i \sum_{p=0}^{Q-1} \frac{-N^p t^{2p+1}}{(2p+1)!} \frac{E(S_N^{2p+1})}{\sigma_N^{2p+1} N^{p+1/2}} + O\left(\frac{t^{2Q+1}}{(2Q+1)!} \frac{E(S_N^{2Q+1})}{\tilde{\sigma}_N^{2Q+1} N^{Q+1/2}}\right) =$$

$$= e^{-t^2/2} + O\left(\frac{t^{2Q}}{2^Q Q!}\right) + O\left(N^{-3/32} \sum_{k=0}^{2Q+1} (CE_N t)^k\right).$$

For sufficiently large T we obtain

$$I := \int_{-t}^{T} \left|\frac{\phi_N(t) - e^{-t^2/2}}{t}\right| dt = O(\frac{T^{2Q}}{2^Q Q!}) + O(N^{-3/32} T^{2Q+1} (CE_N)^{2Q+1})$$

with positive O-constants only depending on α. Choosing $Q = \lfloor \frac{1}{32} \frac{s(N)}{\log_\alpha s(N) + (\log\log\alpha/\log\alpha)} \rfloor$ and $T^2 = Q/2$ yields

$$I \le \omega^Q + O(N^{-\frac{1}{32}} (CE_N)^{2Q+1}) \quad \text{(for some } \omega > 1).$$

Hence we have
(2.8.)

$$\mathrm{prob}\{\frac{S_N}{\tilde{\sigma}_N\sqrt{N}} < y\} = \frac{1}{\sqrt{2\pi}}\int_{-\infty}^{y} e^{-\frac{t^2}{2}}\,dt + O(\frac{1}{\sqrt{Q}}) + O(N^{-\frac{1}{32}}(CE_N)^{2Q+1}) =$$

$$= \frac{1}{\sqrt{2\pi}}\int_{-\infty}^{y} e^{-\frac{t^2}{2}}\,dt + O\left(\sqrt{\frac{\log s(N)}{s(N)}}\right).$$

with an O-constant only depending on α. From (2.8.) our theorem follows immediately. □

Remark 2.9. For constant $s = s(N)$ an analogon of (2.8.) can be proved as a direct application of the results in [6].

Under more restrictive assumptions on $s(N)$ an estimate for $D^{[s(N)]}(\omega_N)$ can be derived for almost all sequences ω; cf. [4].

Theorem 2.10. *Let $s(N)$ be a sequence of positive integers with $s(N) \leq c\log_\alpha N$ and $c < \frac{1}{3}$. Then for $\tau < \frac{1}{3} - c$ and almost all sequences $\omega \in A^\infty$*

$$\alpha^{s(N)}D^{[s(N)]}(\omega_N) \leq N^\tau \text{ for } N \geq N_0.$$

Proof. From Lemma 3 we obtain for every $\varepsilon > 0$

$$\mathrm{prob}\{\omega : \alpha^{s(N)}D^{[s(N)]}(\omega_N) \geq \varepsilon\} = O\left(\frac{N^{2c-1}}{\varepsilon^2}\right).$$

Set $\varepsilon = N^{-\sigma}$ with $0 < \sigma < \frac{1}{2}$ and $a > \frac{1}{-2\sigma+1}$ and obtain by the Borel-Cantelli Lemma that for almost all $\omega \in A^\infty$

$$(k^a - s(k^a) + 1)D^{[s(k^a)]}(\omega_{k^a}) \leq k^{-a\sigma}(k^a - s(k^a) + 1)\alpha^{-s(k^a)} \quad (k \geq k_0).$$

Observing that

$$D^{[s(N)]}(\omega_N)(N - s(N) + 1) \leq P^{[s(k^a)]}(\omega_{k^a})(k^a - s(k^a) + 1) + O(k^{a-1})$$

for $k^a \leq N < (k+1)^a$ yields for almost all $\omega \in A^\infty$

$$D^{[s(N)]}(\omega_N) = O(N^{-\sigma} + N^{-\frac{1}{a}}) = O(N^{-\sigma} + N^{2\sigma-1}).$$

Multiplying with $\alpha^{s(N)}$ gives

$$\alpha^{s(N)}D^{[s(N)]}(\omega_N) = O(N^{-\sigma+c} + N^{2\sigma-1+c})$$

and the result follows immediately. □

Final Remark. In a forthcoming paper of the authors Theorem 2.10. is improved such that almost all ω are $[s(N)]$-u.d, if $s(N) \leq c\log_a N$ with $c < 1$.

References

[1] Goldstern M., Vollständige Gleichverteilung in disketen Räumen, *In Zahlentheoretische Analysis II. (ed. E. Hlawka)* Springer Verlag, Lecture Notes in Math. **1262**, 46-49 (1987).

[2] Hlawka E., Gleichverteilung und Mathematische Linguistik, *Österr. Akad. Wiss. Math. Naturwiss. Kl.* SB II **189**, 209-248 (1980).

[3] Kirschenhofer P., Tichy R.F., Gleichverteilung in diskreten Raumen, In: *Zahlentheoretische Analysis* (E. Hlawka ed.) Springer Verlag, Lecture Notes in Math. **1114**, 66-76 (1985).

[4] Kirschenhofer P., Tichy R.F., Some distribution properties of 0,1-sequences, *Manuscripta Mathematica* **54**, 205-219 (1985).

[5] Knuth D.E.: *The art of computer programming. Vol 2: Seminumerical Algorithms. Reading.* Addison Wesley, 1981.

[6] Philipp W., Ein zentraler Grenzwertsatz mit Anwendungen auf die Zahlentheorie, *Z. Wahrscheinlichkeitstheorie verw. Geb.* **8**, 185-203 (1967).

Ph. Flajolet
INRIA
Rocquencourt, France

P. Kirschenhofer
Technical University
of Vienna
A-1040 Vienna, Austria

R.F. Tichy
Technical University
of Vienna
A-1040 Vienna, Austria

6. On the Distribution of Monochromatic Configurations

P. Frankl, R.L. Graham and V. Rödl

0. Introduction

Much of Ramsey theory is concerned with the study of structure which is preserved under finite partitions, (eg., see [8], [9], [12]). Some of the earliest results in the field were the following.

Schur's Theorem (1916) [17]. *For any partition of the set* $\mathbf{N}$ *of positive integers into finitely many classes, say* $\mathbf{N} = C_1 \cup \ldots \cup C_r$, *some* C_i *must contain a set of the form* $\{x, y, x+y\}$.

Van der Waerden's Theorem (1927) [19]. *For any finite partition of* $\mathbf{N} = C_1 \cup \ldots \cup C_r$, *some* C_i *must contain arbitrarily long arithmetic progressions.*

Ramsey's Theorem (1930) [16]. *For any finite partition of the set* $\binom{\mathbf{N}}{k}$ *of* k*-element subsets of* $\mathbf{N}$, *say* $\binom{\mathbf{N}}{k} = C_1 \cup \ldots \cup C_r$, *some* C_i *must contain the set* $\binom{X}{k}$ *of all the* k*-element sets of some infinite set* $X \subseteq \mathbf{N}$.

It is common in Ramsey theory to call the classes "colors", the partition into r classes an "r-coloring", and the objects belonging to a single class "monochromatic" (cf. [8], [9]).

Each of these results in fact enjoys a "finite" form, which measures the onset of monochromaticity. We abbreviate the interval $\{1, 2, \ldots, N\}$ by $[N]$.

Schur's Theorem (finite form). *For all* $r \in \mathbf{N}$ *there is a least integer* $Sc(r)$ *such that in any* r*-coloring of* $[Sc(r)]$ *there is a monochromatic set of the form* $\{x, y, x+y\}$.

Van der Waerden's Theorem (finite form). *For any* k *and* r *in* $\mathbf{N}$ *there is a least integer* $W(k, r)$ *such that in any* r*-coloring of* $[W(k, r)]$ *there is a monochromatic* k*-term arithmetic progression.*

Ramsey's Theorem (finite form). *For any* k, l *and* r *in* $\mathbf{N}$ *there is a least integer* $R = R(k, l; r)$ *such that in any* r*-coloring of* $\binom{[R]}{k}$ *there is an* l*-element set* $X \subseteq [R]$ *with* $\binom{X}{k}$ *monochromatic.*

The determination of the true orders of growth of the functions $Sc(r)$, $W(k,r)$ and $R(k,l;r)$ are among the most difficult problems in combinatorics. Indeed, each new factor of $\log\log n$ (or even 2!) is usually considered a significant achievement in this quest. A fairly complete survey of this work (as of the time this article is being written) can be found in [10].

In this article we want to focus on somewhat different quantitative aspects of these partition theorems. In one direction, we will ask not *when* the desired monochromatic structure must occur, but rather *how many* monochromatic structures we must have as the size of the set. We are partitioning tends to infinity, and the parameters k, l and r are fixed (cf. Theorems 1,2 and 3).

In another direction we will investigate a measure of the frequency of occurrence of monochromatic structures first suggested for Schur's theorem by Bergelson [1] (cf. Theorems 4 and 5).

We feel that both types of results can contribute to a deeper understanding of these (and other related) fundamental partition results and their various generalizations.

1. Rado's Theorem

In 1930, R. Rado [15] published a far-reaching generalization of Schur's theorem, which dealt with solution sets to systems of homogeneous linear equations over **Z**. (In fact, this striking work formed the basis of Rado's dissertation written under Schur's direction). To describe his results, we first need some terminology.

For an l by k matrix $A = (a_{ij})$ of integers, denote by $\mathcal{L} = \mathcal{L}(A)$ the system of homogeneous linear equations

$$\sum_{j=1}^{k} a_{ij}x_j = 0, \quad 1 \le i \le l. \tag{1.1}$$

We can abbreviate this by writing

$$A\bar{x} = \bar{0}, \ \bar{x} = \begin{pmatrix} x_1 \\ x_2 \\ \vdots \\ x_k \end{pmatrix} = (x_1, \ldots, x_k)^t \tag{1.2}$$

We say that $\mathcal{L}$ is *partition regular* if for any r-coloring of $\mathbf{N}$, there is always a solution to (1.1) with all x_i having the same color.

The matrix A is said to satisfy the *Columns Condition* if it is possible to re-order the column vectors $\bar{a}_1, \bar{a}_2, \ldots, \bar{a}_k$ so that for some choice of indices $1 \leq k_1 < k_2 < \ldots < k_t = k$, if we set

$$A_i := \sum_{j=k_{i-1}+1}^{k_i} \bar{a}_j$$

then

(i) $A_1 = \bar{0}$;

(ii) For $1 < i < t$, A_i can be expressed as a rational linear combination of $\bar{a}_j$, $1 \leq j \leq k_{i-1}$.

A classical results of Rado asserts the following.

Rado's Theorem ([15], [9]). *The system $A\bar{x} = \bar{0}$ is partition regular if and only if A satisfies the Columns Condition.*

Let us call a set $\mathcal{H} \subseteq \mathbf{N}$ *large* if for any partition regular system $A\bar{x} = 0$ and finite coloring of $\mathcal{H}$, there is always a monochromatic solution to $A\bar{x} = 0$. It was shown by Deuber [3] (settling a conjecture of Rado) that large sets have the following partition property: If $\mathcal{H}$ is large and $\mathcal{H} = \mathcal{H}_1 \cup \ldots \cup \mathcal{H}_r$ then for some i, $\mathcal{H}_i$ is large. We next introduce some notation due to Deuber [3].

Definition. $D(m,p,c,) := \{(\lambda_1, \ldots, \lambda_m) :$ for some $i < m$, $\lambda_j = 0$ for $j < i$, $\lambda_i = c > 0$ and $|\lambda_k| \leq p$ for $k > i\}$

A set $S \subseteq \mathbf{Z}^+$ is called an *(m,p,c)-set* if

$$S = \{\sum_{i=1}^{m} \lambda_i y_i : (\lambda_1, \lambda_2, \ldots, \lambda_m) \in D(m,p,c)\}$$

for some choice of $y_1, y_2, \ldots, y_m > 0$.

As shown by Deuber, sets of solutions for partition regular systems $A\bar{x} = 0$ correspond to subsets of (m,p,c)-sets in the following way. Let A be an l by k matrix satisfying the Columns Condition, and let $A_1, A_2, \ldots, A_t$ be the column vector sums coming from the definition of the Columns Condition. We can assume without loss of generality that A has rank l. Then there exist $k - l$ linearly independent solutions to $A\bar{x} = \bar{0}$ which (by the Columns Condition) have the following form *:

* $\bar{x}^t$ denotes the transpose of $\bar{x}$; we will occasionally omit the t if it is clear from context.

$$\begin{array}{llll} & k_1 & k_2-k_1 & \qquad k_t-k_{t-1} \\ \bar{w}_1 & =(1,1,\dots,1, & 0,0,\dots,0,\dots\dots & 0,0,\dots,0)^t \\ \bar{w}_2 & =(\alpha_{21},\dots,\alpha_{2k_1}, & 1,1,\dots,1,\dots\dots & 0,0,\dots,0)^t \\ \vdots & & & \\ \bar{w}_t & =(\alpha_{t1}, & \dots & \dots\alpha_{tk_{t-1}},1,1,\dots,1)^t \\ \bar{w}_{t+1} & =(\alpha_{t+1,1}, & \dots & \dots\dots \quad \alpha_{t+1,k})^t \\ \vdots & & & \\ \bar{w}_{k-l} & =(\alpha_{k-l,1}, & \dots & \dots\dots \quad \alpha_{k-l,k})^t \end{array}$$

where all the α_{ij} are rational. Multiplying all the entries by a sufficiently large integer c, we obtain linearly independent vectors of the following form:

$$(1.3) \qquad \begin{aligned} \bar{v}_1 &= (c,c,\dots,c,0,\dots,0,\dots,0,\dots,0)^t \\ \bar{v}_2 &= (\beta_{21},\dots,\beta_{2,k_1},c,\dots,c,0,\dots,0)^t \\ &\vdots \\ \bar{v}_{t+1} &= (\beta_{t+1,1},\dots,\beta_{t+1,k})^t \\ &\vdots \\ \bar{v}_{k-l} &= (\beta_{k-l,1},\dots,\beta_{k-l,k})^t \end{aligned}$$

where all entries are integers. Set $p = |\max \beta_{ij}|$. Since every solution to $A\bar{x} = \bar{0}$ can be expressed as a linear combination of the vectors $\bar{v}_1, \bar{v}_2, \dots, \bar{v}_{k-l}$, say,

$$\bar{x} = \sum_{i=1}^{k-l} y_i v_i,$$

then in fact each solution of $A\bar{x} = \bar{0}$ is always a subset of some $(k-l,p,c)$-set, and conversely, as claimed.

We are now ready to give the following quantitative version of Rado's Theorem.

Theorem 1. *Let A be an l by k matrix of rank l which satisfies the Columns Condition. Then for any r there exists $c_r(A) > 0$ such that in any r-coloring of $[N]$, $N > N_0$ there are at least $c_r(A)N^{k-l}$ monochromatic solutions to the partition regular system $A\bar{x} = \bar{0}$.*

If we let $\nu_{\mathcal{L}}(N,r)$ denote the minimum possible number of monochromatic solution sets to a system $\mathcal{L}$ whenever $[N]$ is r-colored (so that $\nu_{\mathcal{L}}(N) = \nu_{\mathcal{L}}(N,1)$), then we have as an immediate consequence:

Corollary 1. *If $\mathcal{L}$ is partition regular then for any r there exists $c_r(\mathcal{L}) > 0$ so that*

$$\liminf_{N\to\infty} \frac{\nu_{\mathcal{L}}(N,r)}{\nu_{\mathcal{L}}(N)} \geq c_r(\mathcal{L})$$

Proof of Theorem 1. The proof will use the following version of Deuber's theorem ([3]).

Theorem. *For every choice of m, p, c and r there exist M, P and C such that for any r-coloring of*

$$J = \{\sum_{i=1}^{M} \lambda_i Y_i : (\lambda_1, \ldots, \lambda_M) \in D(M, P, C)\}$$

there exist pairwise disjoint sets $B_1, B_2, \ldots, B_m \subseteq [m]$ and

$$y_i = \sum_{j \in B_i} \xi_j Y_j, \quad 1 \leq |\xi_j| \leq P,\ 1 \leq i \leq m,$$

such that all linear combinations

$$\sum_{i=1}^{m} \lambda_i y_i, \quad (\lambda_1, \lambda_2, \ldots, \lambda_m) \in D(m, p, c)$$

are monochromatic.

Now, given our l by k matrix A of rank l satisfying the Columns Condition, we know by the preceding remarks that the entries of the set of solution vectors of $A\bar{x} = \bar{0}$ all belong to some $(k-l, p, c)$-set. Set $m = k - l$ and let M, P and C be the integers from Deuber's theorem. Choose $N \gg M$ to be very large. Consider all the M-tuples $(Y_1, Y_2, \ldots, Y_M)$ of integers Y_i satisfying

$$(1.4) \qquad 0 < Y_i \leq \frac{N}{MC} \text{ and } Y_i \equiv (2P+1)^i \ mod (2P+1)^M$$

for $1 \leq i \leq M$. There are at least $c_1 N^M$ such M-tuples for some constant $c_1 > 0$ not depending on N. For such an M-tuple $(Y_1, Y_2, \ldots, Y_M)$, consider the (M, P, C)-set

$$J(Y_1, \ldots, Y_M) = \{\sum_{i=1}^{M} \lambda_i Y_i : (\lambda_1, \ldots, \lambda_M) \in D(M, P, C)\}$$

Let $[N] = C_1 \cup \ldots \cup C_r$ be an r-coloring of $[N]$. By Deuber's theorem we can find disjoint subsets $B_1, \ldots, B_{k-l} \subseteq [M]$ and $y_i = \sum_{j \in B_i} \xi_j Y_j$, $1 \leq |\xi_j| \leq P$, so that all the linear combinations $\sum_{i=1}^{m} \lambda_i y_i$ $(\lambda_1, \lambda_2, \ldots, \lambda_m) \in D(m, p, c)$, have the same color. In particular, $\bar{x} = \sum_{i=1}^{k-l} y_i \bar{u}_i$ (from (1.3)) is a monochromatic solution to the system $A\bar{x} = \bar{0}$. This therefore gives, with multiplicity, at least $c_1 N^M$ monochromatic solutions (one for each choice of $(Y_1, \ldots, Y_M)$). Our proof will be complete if we can show that each of these solutions can occur at most $N^{M-(k-l)}$ times.

To see this, suppose $(x_1, \ldots, x_k)$ is some solution obtained above, i.e., for some choice of $(y_1, \ldots, y_{k-l})$, the x_i are *fixed* linear combinations of the y_i. Then, we must show that the same monochromatic (m,p,c)-set is obtained at most $N^{M-(k-l)}$ times. However, given y_i, its residue modulo $(2P+1)^M$ uniquely determines the λ_j, $1 \le j \le M$ from (1.4). Thus, the possible $Y_1, \ldots, Y_M$ must satisfy $k-l$ linear equations which involve pairwise disjoint sets of unknowns among them. This gives the required bound and the proof is complete. □

2. Van der Waerden's Theorem

A natural question raised in connection with van der Waerden's Theorem by Erdős and Turán over 50 years ago was that of identifying which of the color classes must contain the desired arbitrarily long arithmetic progressions. In particular, they conjectured that the "largest" color class should always have this property. To make this precise, for a set $X \subseteq \mathbf{N}$, define the upper density $\bar{d}(X)$ of X by:

$$\bar{d}(X) := \limsup_{n\to\infty} \frac{|X \cap [n]|}{n}$$

In 1975, Szemerédi finally settled the conjecture of Erdős and Turán, by proving the following celebrated result.

Theorem of Szemerédi [18]. *If $X \subseteq \mathbf{N}$ satisfies $\bar{d}(X) > 0$ then X contains arbitrarily long arithmetic progressions.*

This result of course implies van der Waerden's Theorem, and it was, in fact, hoped that it might lead to improved estimates for $W(k,r)$. This did not happen (yet) though since Szemerédi's proof in fact uses van der Waerden's Theorem. More recently, Furstenberg and Katznelson [6], [7] have given alternate proofs and generalizations of Szemerédi's Theorem using techniques from ergodic theory and topological dynamics (which however, do not shed any light on the true values of $W(k,r)$).

Observe that a k-term arithmetic progression $(a, a+d, a+2d, \ldots, a+(k-l)d)$ can be viewed as a solution $\bar{x} = (x_1, x_2, \ldots, x_k)$ to the system of equations (over $\mathbf{N}$)

$$x_2 - x_1 = x_3 - x_2 = \ldots = x_k - x_{k-1} \neq 0$$

In this section we establish the density analogue to Theorem 1 for the appropriate systems of linear equations.

The system

$$A\bar{x} = \bar{0} \tag{2.1}$$

is said to be *density regular* if for any set $X \subseteq \mathbf{N}$ of positive upper density there is a vector $\bar{x}$ satisfying (2.1) and having all entries belonging to X.

If it happens that (2.1) has the vector $\bar{x} = \bar{1} = (1, 1, \ldots, 1)$ as a solution then, of course, for any $m \in \mathbf{N}$, $\bar{x} = m \cdot \bar{1} = (m, m, \ldots, m)$ is also a solution. In this case, (2.1) is trivially density regular. However, the solution $m \cdot \bar{1}$ is normally not considered to be very interesting. For example, for the density regular system

$$x_1 - 2x_2 + x_3 = 0$$

the solutions (x_1, x_2, x_3) are just the 3-term arithmetic progressions, provided the x_i are distinct.

With these considerations in mind, let us call the system (2.1) *irredundant*, if (2.1) *does not* imply that $x_i = x_j$ for $i \neq j$. Also, let us call a solution $\bar{x} = (x_1, \ldots, x_k)$ to (2.1) *proper* if all the x_i are distinct.

Fact 2.1. *If $A\bar{x} = \bar{0}$ is irredundant then it has a proper solution.*
Proof. For each choice of $i < j$, let $\bar{x}^{(ij)} = (x_1^{(ij)}, x_2^{(ij)}, \ldots, x_k^{(ij)})$ be a solution to (2.1) with $x_i^{(ij)} \neq x_j^{(ij)}$, which exists by hypothesis. Thus, for any integer N, $\bar{x}^* = (x_1^*, x_2^*, \ldots, x_k^*)$ with

$$x_t^* = \sum_{i<j} N^{ki+j} x_t^{(ij)}$$

is also a solution to (2.1) by linearity. However, if $N > \max_{i,j,t}(x_t^{(ij)})$ then all x_t^* are distinct. □

Fact 2.2. *An irredundant system $A\bar{x} = \bar{0}$ has a proper solution in every set X of positive upper density if and only if $A \cdot \bar{1} = 0$.*
Proof. First, since X has positive upper density then by Szemerédi's Theorem, X contains arbitrarily long arithmetic progressions. Suppose $\bar{x}_0 = (b_1, b_2, \ldots, b_k)$ is a proper solution of $A\bar{x}_0 = \bar{0}$, i.e., all the b_k are distinct. Let $B := \max_k b_k$ and let $P = \{c + \lambda d : \lambda \in [B]\}$ be a B-term arithmetic progression in X. If $\bar{1}$ also satisfies $A \cdot \bar{1} = \bar{0}$ then so does the linear combination

$$\bar{x}^* = c \cdot \bar{1} + d\bar{x}_0 = (c + b_1 d, c + b_2 d, \ldots, c + b_n d)$$

which is proper, and furthermore, has all entries in $P \subseteq X$, as desired.

In the other direction, suppose $A\bar{x} = \bar{0}$ has a proper solution in every set of positive upper density. Let $N > \sum_{i,j} |a_{ij}|$ where a_{ij} ranges over all entries of A. Consider the set $Y = \{Ny + 1 : y \in \mathbf{N}^+\}$ with (upper) density $1/N$. Suppose $\bar{x} = (x_1, \ldots, x_n)$ satisfies $A\bar{x} = \bar{0}$ where each $x_k = Ny_k + 1 \in Y$. Thus,

$$0 = \sum_j a_{ij} x_j = \sum_j a_{ij}(Ny_j + 1) = N \sum_j a_{ij} y_j + \sum_j a_{ij}$$

for $1 \leq i \leq m$. By the choice of N, this implies that $\sum_j a_{ij} = 0$ for all i. This is exactly the statement that $A\bar{1} = \bar{0}$, as required. This completes the proof. □

Theorem 2. *Let A be an l by k matrix of rank l so that $A\bar{x} = 0$ is irredundant and $A\bar{1} = \bar{0}$. Then for any $\epsilon > 0$ there is a constant $c_\epsilon = c_\epsilon(A) > 0$ so that if $N > N_0(A, \epsilon)$ and $X \subseteq [N]$ with $|X| > \epsilon N$ then X must contain at least $c_\epsilon N^{k-l}$ proper solutions $\bar{x}$ to $A\bar{x} = \bar{0}$.*

Proof. Let $\epsilon > 0$ be arbitrary (but fixed) and let $X \subseteq [N]$ with $|X| > \epsilon N$ be given, where it will be useful to think of N as being very large. Since A has rank l, the space of all (rational) solutions $\bar{x}$ to $A\bar{x} = \bar{0}$ has dimension $k - l$. Let $\bar{v}_0 = \bar{1}$, $\bar{v}_1, \ldots, \bar{v}_m$ be linearly independent integer solutions to $A\bar{x} = \bar{0}$ where $m := k - l - 1$ and for $\bar{v}_i = (v_{i1}, \ldots, v_{ik})^t$, we can assume without loss of generality, all $v_{ij} \geq 0$ (since if not, then we can repeatedly add $\bar{1}$ to $\bar{v}_i$ until this is true). Define $t := 1 + \max_{ij} v_{ij}$.

For $u \in \mathbf{N}$ and each vector $\bar{y} = (y_1, \ldots, y_m)$ with $y_i \in \mathbf{N}$, define the m-box $B_u(\bar{y})$ to be the set

$$\{(a_1 y_1, a_2 y_2, \ldots, a_m y_m) : 0 \leq a_i < u,\ 1 \leq i \leq m\}$$

Further, define the projection $\pi : B_u(\bar{y}) \to \mathbf{Z}$ by

$$\pi[(a_1 y_1, \ldots, a_m y_m)] = \sum_{i=1}^{m} a_i y_i$$

By a theorem of Furstenberg and Katznelson ([6], [7]), there is an integer T so that for any $\bar{Y} = (Y_1, \ldots, Y_m)$ with $Y_i \in \mathbf{N}^+$, if $X^* \subseteq B_T(\bar{Y})$ with $|X^* > \frac{\epsilon}{2}|B_T(\bar{Y})| = \frac{\epsilon}{2}T^m$ then there exists a "translated" m-box $\bar{A} + B_t(A_0\bar{Y}) \subseteq X^*$, where $\bar{A} = (A_1 Y_1, \ldots, A_m Y_m)$ and $A_0, A_1, \ldots, A_m \in \mathbf{N}^+$.

Now, consider the set of all integer vectors $\bar{Y} = (Y_1, Y_2, \ldots, Y_m)$ which satisfy the following constraints:

(i) $0 \leq Y_i < \epsilon^2 N / mT$, $1 \leq i \leq m$;

(ii) $Y_i \equiv T^{i-1} (\mathrm{mod}\ T^m)$, $1 \leq i \leq m$.

Note that if $\bar{P} = (a_1 Y_1, \ldots, a_m Y_m) \in B_T(\bar{Y})$, $\bar{P}' = (a'_1 Y_1, \ldots, a'_m Y_m) \in B_T(\bar{Y})$ and $\pi(\bar{P}) = \pi(\bar{P}')$ then by (ii)

$$\sum_{i=1}^{m} a_i T^{i-1} \equiv \sum_{i=1}^{m} a'_i T^{i-1} \pmod{T^m}$$

which in turn implies $a_i = a'_i$ for all i, since $0 \leq a_i,\ a'_i < T$. Thus, π is 1-to-1 on $B_T(\bar{Y})$. Also, by (i)

$$0 \leq \pi(\bar{P}) < \epsilon^2 N$$

Let us call an integer $\underline{a} \in [N]$ "good" if

$$B(a) := a + \pi(B_T(\bar{Y})) \subseteq [N]$$

and

$$|X \cap B(a)| > \frac{\epsilon}{2} T^m$$

It is easy to see that for a fixed constant $\delta = \delta(\epsilon) > 0$, the set $A = \{a \in [N] : a \text{ is good}\}$ satisfies

$$|A| > \delta N.$$

By the choice of T, for each $a \in A$, $X \cap B(a)$ contains the translated projection

$$Y_0 + \pi(B_t(A_0\bar{Y}))$$

for some $Y_0, A_0 \in \mathbf{N}$. Furthermore, by the choice of t, this in turn contains all components of the solution

$$\bar{x} = Y_0 \cdot \bar{1} + \sum_{i=1}^{m} A_0 Y_i \bar{v}_i$$

to $A\bar{x} = \bar{0}$. Since there are cN^{m+1} ways to choose the $Y_0, Y_1, \ldots, Y_m$ for a positive constant c (depending on ϵ and A) then the theorem will be proved if we can show that no solution $\bar{x}$ to $A\bar{x} = \bar{0}$ can arise this way in more than a bounded number of ways.

To see this, first note that since the $(k-l)$ by k matrix $V = (v_{ij})$ formed from the (linearly independent) solution vectors $\bar{v}_i$, $0 \le i < k-l$, has rank $k-l$ then we can assume without loss of generality (by relabelling, if necessary) that the $(k-l)$ by $(k-l)$ submatrix $V' = (v_{ij})_{0 \le i,j < k-l}$ is non-singular. Suppose $\bar{x} = (x_1, \ldots, x_k)$ has all its components x_i lying in some set $Y_0 + \pi(B_T(\bar{Y})) \subseteq [N]$ where $\bar{Y} = (Y_1, \ldots, Y_m)$ satisfies (i) and (ii). For each of the $k!$ permutations σ on $[k]$, consider the vector $\bar{x}_\sigma = (x_{\sigma(1)}, \ldots, x_{\sigma(k)})$. If

$$\bar{x}_\sigma = \sum_{i=0}^{m} Y_i \bar{v}_i$$

then by the non-singularity of V', the first $k-l$ coordinates of $\bar{x}_\sigma$ determine all the Y_i. Thus, each such $\bar{x}$ can arise from at most $k!$ choices for the Y_i.

Finally, we observe that almost all of these $c'N^{k-l}$ solutions $\bar{x}$ to $A\bar{x} = \bar{0}$ are proper solutions. This is because, by hypothesis, for $i \ne j$, the space of solutions $\bar{x}$ with $x_i = x_j$ corresponds to a non-trivial dependence between the coefficients Y_i, $1 \le i \le m$, resulting in at most $O(N^{k-l-1})$ such solutions.

This completes the proof of the theorem. □

Let $v^*_{\mathcal{L}}(N;\epsilon)$ denote the minimum possible number of proper solutions to a system $\mathcal{L} = \mathcal{L}(A)$ which can belong to a set $X \subseteq [N]$ having $|X| > \epsilon N$. The following corollary is immediate.

Corollary. *If A is irredundant and $\mathcal{L} = \mathcal{L}(A)$ is density regular (i.e., $A \cdot \bar{1} = \bar{0}$) then for any $\epsilon > 0$ there exists $c^*_\epsilon(\mathcal{L}) > 0$ such that*

$$\liminf_{N \to \infty} \frac{v^*_{\mathcal{L}}(N;\epsilon)}{v_{\mathcal{L}}(N)} \ge c^*_\epsilon(\mathcal{L}),$$

where $v_{\mathcal{L}}(N)$ denotes the total number of solutions $\mathcal{L}$ has in $[N]$.

When the Corollary is applied to the system

$$\mathcal{L}^* : x_2 - x_1 = x_3 - x_2 = \ldots = x_k - x_{k-1}$$

we obtain the desired qualitative form of Szemerédi's Theorem, namely for some $c = c_\epsilon^*(\mathcal{L}^*) > 0$, if $N > N_0$ then

$$v_{\mathcal{L}^*}^*(N;\epsilon) \geq cN^2,$$

which is, of course, up to the value of c, the best one could hope for here.

3. Ramsey's Theorem

It turns out the analogues of the preceding results for Ramsey's Theorem follow rather easily from an averaging argument. We sketch it here for completeness.

Theorem 3. *For all k, l and r in* N *there exists $c = c(k,l,r) > 0$ such that for any r-coloring of $\binom{[N]}{k}$, $N \geq R(k,l;r)$, there are at least $c\binom{N}{l}$ l-sets $Y \subseteq [N]$ for which $\binom{Y}{k}$ is monochromatic.*

Proof. Let $R = R(k,l;r)$, and suppose $\binom{[N]}{k}$ is arbitrarily r-colored with $N \geq R$. Then for any R-set $Z \in \binom{[N]}{R}$ there is always some l-set $Y \in \binom{Z}{l}$ with $\binom{Y}{k}$ monochromatic. Call such an l-set Y "good". Now each good Y can occur in at most $\binom{N-l}{R-l}$ different $Z \in \binom{[N]}{R}$. Since there are $\binom{N}{R}$ different $Z \in \binom{[N]}{R}$ then there must be at least

$$\frac{\binom{N}{R}}{\binom{N-l}{R-l}} > c_R\binom{N}{l}$$

good sets $Y \in \binom{[N]}{l}$, i.e., such that $\binom{Y}{k}$ is monochromatic. □

Note that as in Theorems 1 and 2, a positive proportion of the objects under consideration is guaranteed to be monochromatic. This phenomenon does not always occur, however, as the following example shows. It is known (see [8]) that for each r there is an $F(r)$ so that for any r-coloring of all the subsets of $[F(r)]$ we can always find nonempty disjoint sets $A, B \subseteq [F(r)]$ so that A, B and $A \cup B$ all have the same color. Now, for a fixed N, consider the 2-coloring $\mathcal{X}$ of the subsets of $[N]$ given by:

$$\mathcal{X}(X) = \begin{cases} 0, & \text{if } |X| \leq N/2, \\ 1, & \text{if } |X| > N/2. \end{cases}$$

However, with this coloring there are only $O(2^{3N/2})$ monochromatic triples $\{A, B, A \cup B\}$ while in $[N]$ there are $(1 + o(1))3^N$. Thus, we do not get a positive proportion in this case.

4. An Iterated Density Theorem for the Strong van der Waerden Theorem

The following strengthening of van der Waerden's Theorem was used by Rado [15] in his work on partition regular systems.

Strong van der Waerden Theorem. *In any finite coloring of* N *there must exist for all* $p \in \mathbf{N}$, *a monochromatic set of the form*

$$\{x, y, x+y, 2x+y, \ldots, (p-1)x+y\}.$$

Note that this set consists of a p-term arithmetic progression together with its common difference. For the special case of $p = 2$, this reduces to the set $\{x, y, x+y\}$ which occurs in Schur's Theorem. However, even in this case it is clear that there is no direct density analogue to this theorem (as Szemerédi's Theorem was for the ordinary van der Waerden Theorem) since, for example, the set of odd integers $\{2k+1 : k \in \mathbf{N}\}$ has (upper) density $1/2$ and yet contains *no* set of the form $x, y, x+y$. Nevertheless, it is possible to prove a result which asserts that in any finite partition of N, there are "many" monochromatic sets of the form $\{x, y, x+y\}$. This was first done by Bergelson, who recently proved the following.

Theorem 1. *In any finite coloring of* N, *we have*

$$\bar{d}\{x : \bar{d}\{y : \{x, y, x+y\} \text{ is monochromatic} > 0\} > 0\} \tag{4.1}$$

Actually, Bergelson proves the following somewhat stronger result which does not, however, guarantee that either set has upper density bounded away from 0 as a function only of the number of colors.

Bergelson's Theorem (1986) [1]. *In any finite coloring of* N, *there is always some color class* C *with* $\bar{d}(C) > 0$ *such that for any* $\epsilon > 0$,

$$\bar{d}\{x : \bar{d}\{y : x, y, x+y \text{ is monochromatic }\} \geq \bar{d}(c)^2 - \epsilon\} > 0$$

In this section we will prove an iterated density version of the strong van der Waerden Theorem, which will imply, in particular, a strengthening of (4.1), both in having explicit functions in the lower bounds, and in the replacement of $\bar{d}$ by **d**. We first introduce a slightly modified form of the (m, p, c)–sets introduced in Section 1, called $(m, p, c)'$–sets, which will be useful in what follows.

For m, p and c in $\mathbf{N}$, we mean by an $(m,p,c)'$-set, a subset of $\mathbf{N}$ which can be formed as follows, for suitable $a_1, a_2, \ldots, a_m \in \mathbf{N}$:

$$< a_1, \ldots, a_m > := \{\lambda_1 a_1 + \ldots + \lambda_{i-1} + c a_i : 0 \le \lambda_j < p,\ 1 \le j < i,\ 1 \le i \le m\}.$$

We also define

$$[a_1, \ldots, a_m] := \{\lambda_1 a_1 + \ldots + \lambda_m a_m : 0 \le \lambda_i < p\},$$

where in both cases the values of p and c will be understood from the context if only the left-hand sides are used.

Thus,

$$(4.2) \qquad < a_1, \ldots, a_m > = < a_1, \ldots, a_{m-1} > \cup \{[a_1, \ldots, a_{m-1}] + c a_m\}.$$

Suppose now that $\mathbf{N}$ is arbitrarily r-colored. A basic result of Deuber [3] then implies that there exist $M, P, C \in \mathbf{N}$ so that any $(M,P,C)'$-set must always contain a monochromatic $(m,p,c)'$-set.

Our result will deal with $(2,p,1)'$-sets. These are just sets of the form $< x, y > = \{x, y, x+y, 2x+y, \ldots, (p-1)x+y\}$. In what follows, the integers M, P and C will denote the values needed in Deuber's theorem to force monochromatic $(2,p,1)'$-sets.

Theorem 4. (Iterated Strong van der Waerden Theorem). *In any r-coloring of $\mathbf{N}$ and for any p, there is a $\delta = \delta(r,p) > 0$ such that*

$$(4.3) \qquad \mathbf{d}\{x : \bar{d}\{y :< x, y > \text{ (is monochromatic)}\} > \delta\} > \delta.$$

Proof. We assume $p > 2$ (the case $p = 2$ is very similar, and is omitted). Define

$$B = B(\delta) := \{x : \bar{d}y :< x, y > \text{ is monochromatic} \le \delta\}$$

and let $\bar{B} := \mathbf{N} \backslash B$, the complement of B in $\mathbf{N}$. Assume to the contrary that $\mathbf{d}(\bar{B}) = 1 - \bar{d}(B) \le \delta$. Let $B_1 := \{d \in B : < b > \subseteq B\}$, and let $a_1 \in B_1$ be the least element in B_1. Next, define

$$B_2 = \{b \in B_1 : [a_1] + Cb \subseteq B_1\}$$

Thus, if $b \in B_2$, then

$$[a_1] + Cb \subseteq B_1 \subseteq B \text{ and } < a_1 > \subseteq B$$

so that $< a_1, b > \subseteq B$. Next, select (if possible) $a_2 \in B_2$ so that $< a_1, a_2 >$ contains no monochromatic $(2,p,1)'$-set. Define

$$B_3 := \{b \in B_2 : [a_1, a_2] + Cb \subseteq B_2\}.$$

Thus, for $b \in B_3$, $< a_1, a_2, b > \subseteq B$. Continuing, we select (if possible) $a_3 \in B_3$ so that $< a_1, a_2, a_3 >$ contains no monochromatic $(2, p, 1)'$-set, and we define

$$B_4 := \{b \in B_3 : [a_1, a_2, a_3] + Cb \subseteq B_3\}, \text{ etc.}$$

In general, after

$$B_j := \{b \in B_{j-1} : [a_1, \ldots, a_{j-1}] + Cb \subseteq B_{j-1}\}$$

is formed, we see that for $b \in B_j$, $< a_1, \ldots, a_{j-1}, b > \subseteq B$. We then select (if possible) $a_j \in B_j$ so that $< a_1, \ldots, a_j >$ contains no monochromatic $(2, p, 1)'$-set, and define $B_{j+1} := \{b \in B_j : [a_1, \ldots, a_j] + Cb \subseteq B_j\}$, etc. (where, of course, throughout this construction $< a_1, \ldots, a_i >$ denotes an (i, P, C)-set). By Deuber's theorem this process must terminate with the formation of B_t, for some $t < M$. In order to guarantee that this is actual cause for termination, we need to know that the various $B_j's$ are nonempty. This fact is implied by the following elementary lemma.

Lemma. *Let $A \subseteq \mathbf{N}$, $D \subseteq \mathbf{N}$, $C \in \mathbf{N}$ with A finite, and define*

$$D' = \{d \in D : Cd + A \subseteq D\}$$

Then

$$1 - \bar{d}(D') \le C \mid A \mid (1 - \bar{d}(D)) \tag{4.4}$$

The proof of this result is elementary and will be omitted. Thus, by (4.4) and (4.2) (with C in place of c), we have

$$1 - \bar{d}(B_{j+1}) \le C \mid [a_1, \ldots, a_j] \mid (1 - \bar{d}(B_j)) \le CP^j(1 - \bar{d}(B_j)). \tag{4.5}$$

Consequently,

$$1 - \bar{d}(B_t) \le C^t P^{\binom{t+1}{2}}(1 - \bar{d}(B)) \le \delta C^M P^{\binom{M}{2}} \tag{4.6}$$

so that for δ sufficiently small, all $\bar{d}(B_i)$ are at least 1/2 (say) for $1 \le i \le t$.

Therefore, the set $\{a_1, \ldots, a_{t-1}\}$ has the properties:

(i) $< a_1, \ldots, a_{t-1} >$ contains no monochromatic $(2, p, 1)'$-set;

(ii) For any $b \in B_t$, $< a_1, \ldots, a_{t-1}, b > \subseteq B$ contains some monochromatic $(2, p, 1)'$-set, say $< x(b), y(b) >$.

Thus,

$$x(b) = \lambda_1 a_1 + \ldots + \lambda_{j-1} a_{j-1} + C a_j \tag{4.7}$$

for some $j = j(b) \le t - 11$. To see this, suppose otherwise, i.e., suppose that

$$x(b) = \lambda_1 a_1 + \ldots + \lambda_{t-1} a_{t-1} + Cb.$$

Since $p > 2$ by hypothesis then $< x(b), y(b) >$ contains the element $2x(b) + y(b) > 2Cb$. However, this is impossible if b is sufficiently large since we have assumed

$$< x(b), y(b) > \subseteq < a_1, \ldots, a_{t-1}, b >$$

and the largest element of $< a_1, \ldots, a_{t-1}, b >$ is less than $(a_1 + \ldots + a_{t_1})p + Cb$.

On the other hand, we must have

$$y(b) = \lambda'_1 a_1 + \ldots + \lambda'_{t-1} a_{t-1} + Cb \tag{4.8}$$

since, if not, say $y(b) = \lambda'_1 a_1 + \ldots + \lambda'_{j-1} a_{j-1} + Ca_j$ for some $j < t$, then no element of $< x(b), y(b) >$ is large enough to use b, so that $< a_1, \ldots, a_{t-1} >$ must have already contained $< x(b), y(b) >$, which is a contradiction.

Now, for each $b \in B_t$, there are fewer than MP^{2M} choices for $j(b)$, $\bar{\lambda}(b) = (\lambda_1, \ldots)$ and $\bar{\lambda}'(b) = (\lambda'_1, \ldots)$. Hence, for some choice of $j_0(b), \lambda_0(b)$ and $\bar{\lambda}'_0(b)$, the set of (large) $b \in B_t$ with $j(b) = j_0(b)$, $\bar{\lambda}(b) = \bar{\lambda}_0(b)$ and $\bar{\lambda}'(b) = \bar{\lambda}'_0(b)$ has upper density at least $\bar{d}(B_t)/MP^{2M}$. Call this set B^*. Also

$$\bar{d}\{y : y = y(b) \text{ for } b \in B^*\} \geq \frac{1}{C}\bar{d}(B^*) \geq \bar{d}(B_t)/CMP^{2M} \tag{4.9}$$

since $y = y(b) = \lambda'_1 a_1 + \ldots + \lambda'_{t-1} a_{t-1} + Cb$. Note that

$$x(b) = \lambda_1 a_1 + \ldots + \lambda_{j0} a_{j0} \in < a_1, \ldots, a_{t-1} > \subseteq B.$$

Therefore, $\bar{d}\{y(b) : b \in B^*\} \leq \delta$ since $< x(b), y(b) >$ is monochromatic. This implies

$$\bar{d}(B_t)/CMP^{2M} \leq \delta,$$

i.e.,

$$\bar{d}(B_t) \leq \delta MP^{2M}C \tag{4.10}$$

However, this contradicts (4.6) if δ is sufficiently small. Hence, the initial assumption that $\mathrm{d}(\bar{B}) \leq \delta$ is untenable, and (4.3) must hold. This completes the proof. □

Note that this proof shows that δ can be chosen to be $(M^2 C^{M+1} P^{2M^2})^{-1}$, for example.

5. An Iterated Density Ramsey Theorem

The obvious density version of the finite form of Ramsey's theorem is clearly false, as can be seen, for example, by considering the complete bipartite graph $K_{n,n}$. This graph has more than half the possible number of edges for a graph with $2n$ vertices but contains no triangle. However, there is an iterated density version (in the spirit of Theorem 4) which is valid. This we now give.

Fix $k \le l$ and r, and suppose $\binom{N}{k}$ is r-colored. For $a_1 < \ldots < a_{l-1}$, define

$$\Gamma(a_1, \ldots, a_{l-1}) := \{x : \binom{a_1, \ldots, a_{l-1}, x}{k} \text{ is monochromatic}\},$$

and for $0 \le i < l-1$,

$$\Gamma(a_1, \ldots, a_i) := \{x : \bar{d}(\Gamma(a_1, \ldots, a_i, x)) > \delta\} \tag{5.1}$$

where $\delta = \delta(k, l, r) = 2^{-R}$ and $R := (k, l; r)$, the ordinary Ramsey number. For $i = 0$, we denote the expression in (5.1) by Γ.

Theorem 5. *For all $k \le l$ and r,*

$$\mathrm{d}(\Gamma) > \delta. \tag{5.2}$$

Proof. Assume $\mathrm{d}(\Gamma) \le \delta$. Thus

$$\bar{d}(\mathrm{N} \backslash \Gamma) \ge 1 - \delta.$$

Note that

$$x \notin \Gamma(a_1, \ldots, a_i) \Rightarrow \bar{d}(\Gamma(a_1, \ldots, a_i, x)) \le \delta.$$

Define S_i, $i = 1, 2, \ldots$, as follows:

$$S_1 = \{s_1\} \text{ where } s_1 \in \mathrm{N} \backslash \Gamma \text{ is arbitrary.}$$

Suppose $S_j = \{s_1, s_2, \ldots, s_j\}$ has been defined. Form $S_{j+1} = S_j \cup \{s_{j+1}\}$ by choosing s_{j+1} (if possible) so that:

(i) $s_{j+1} \in \mathbf{N} - \cup_{u=0}^{l-1} \cup_{1 \le i_1 < \ldots < i_u \le j} \Gamma(S_{i_1}, \ldots, S_{i_u}) := C_j$,

(ii) No $Y \in \binom{S_{j+1}}{l}$ has $\binom{Y}{k}$ monochromatic.

Note that since

$$\bar{d}(C_j) \ge 1 - \{\binom{j}{0} + \binom{j}{1} + \ldots + \binom{j}{k-1}\}\delta \ge 1 - 2^j \delta$$

then we never get stuck because of (i). However, by Ramsey's Theorem, we must eventually halt because of condition (ii), say with the formation of S_t, for some $t < R$. By the definition of S_t, for each $c \in C_t$, there is a set $X(c) \in \binom{S_t}{l-1}$ such that $\binom{X(c) \cup c}{k}$ is monochromatic. Thus, there exists a set $X_0 = \{s_{j_1} < \ldots < s_{j_{l-1}}\} \in \binom{S_t}{l-1}$ such that

(5.3)

$$\bar{d}\{c \in C_t : X(c) = X_0\} \ge \bar{d}(C_t) / \binom{t}{l-1} \ge (1 - 2^t \delta) / \binom{t}{l-1} > 2^{-R} = \delta$$

by the choice of δ. However, by construction,

$$s_{j_{l-1}} \notin \Gamma(s_{j_1}, \ldots, s_{j_{l-2}})$$
$$\Rightarrow \bar{d}(\Gamma(s_{j_1}, \ldots, s_{j_{l-1}})) \le \delta$$

$$\Rightarrow \bar{d}\{u : \binom{X_0 \cup u}{l} \text{ is monochromatic } \} \le \delta$$

which contradicts (5.3). This proves the theorem. □

6. Concluding Remarks

There are a number of other examples known for which some of the preceding extensions can be proved (cf. [4], [5]). These include several of the canonical partition theorems, in which an arbitrary number of colors can be used (but a wider class of colored objects is allowed); see [2], [11], [13], [14]. However, we have barely scratched the surface for what might be looked at here. For example, if $<< x,y,z >>:= \{x,y,z,x+y,x+z,y+z,x+y+z\}$ then is it true that for any r-coloring of N there is a $\delta > 0$ such that

$$\mathrm{d}\{x : \bar{d}\{y : \bar{d}\{z :<< x,y,z >> \text{ is monochromatic } \} > \delta\} > \delta\} > \delta?$$

We also point out that we have very little idea as to the true values of the various constants ($\delta's$ and $c's$) appearing in our theorems. Of course, since these typically depend on the corresponding values of the classical Ramsey numbers $Sc(r)$, $W(k,r)$ and $R(k,l;r)$ which themselves are far from being completely understood, (see [10]) then it is not surprising that our current knowledge here is very incomplete.

References

[1] V.Bergelson, A density statement generalizing Schur's theorem, *J.Combin. Th. (A)* **43** (1986), 338-343.

[2] W.Deuber, R.L.Graham, H.J.Prőmel, B.Voigt, A Canonical Partition Theorem for Equivalence Relations on 2^t, *J.Combinatorial Th. (A)*, **34**, (1983), 331-339.

[3] W.Deuber, Partitionen und lineare Gleichungsysteme, *Math. Z.* **133** (1973), 109-123.

[4] P.Frankl, R.L.Graham and V.Rödl, Quantitative theorems for regular systems of equations, (to appear).

[5] P.Frankl, R.L.Graham and V.Rödl, Iterated combinatorial density theorems, (to appear).

[6] H.Furstenberg, Y.Katznelson, An ergodic Szemerédi theorem for commuting transformations, *J.Analyse Math.* **34** (1978), 275-291.

[7] H.Furstenberg, Y.Katznelson and D.Ornstein, The ergodic theoretic proof of Szemerédi's Theorem, *Bull Amer. Math. Soc.* **7** (1982), 527-552.

[8] R.L.Graham, Rudiments of Ramsey Theory, *Regional Conference Series in Mathematics*, no. 45, AMS, 1981.

[9] R.L.Graham, B.L.Rothschild, J.H.Spencer, *Ramsey Theory*, John Wiley & Sons, Inc., 1980.

[10] R.L.Graham and V.Rödl, Numbers in Ramsey theory, *in Surveys in Combinatorica 1987* (C. Whitehead, ed) LMS Lecture Note Series 123, Cambridge, 1987, 111–113.

[11] H.Lefmann, Kanonische Partitionssatze, *Ph.D. Dissertation*, Univ. of Bielefeld (1985).

[12] J. Nešetřil, V.Rödl, *Partition Theory and its Applications*, Surveys in Combinatorics 1979 (B. Bollobás, ed) LMS Lecture Note Series 38, Cambridge, 1979, 96–157.

[13] H.J.Prömel, V.Rödl, An elementary proof of the canonizing version of Gallai-Witt's Theorem, *J.Combinatorial Theory (A)*, **42**, 2 (1986), 144-149.

[14] H.J.Prömel, B.Voigt, Canonical Partition Theorems for Parameter Sets, *J. Combinatorial Theory (A)*, **35** (1985), 309-327.

[15] R.Rado, Studien zur Kombinatorik, *Math. Z.* **36** (1933), 242-280.

[16] F.P.Ramsey, On a problem in formal logic, *Proc. London Math. Soc.* (2) **30** (1930), 264-285.

[17] I.Schur, Über die Kongruenz $x^m + y^m = z^m \pmod p$, *Jber.Deutsch. Math. Verein.* **25** (1916), 114-116.

[18] E.Szemerédi, On Sets of Integers containing no k Elements in Arithmetic Progression, *Acta Arith.* **27** (1975), 199-245.

[19] B.L. van der Waerden, Beweis einer Baudetschen Vermutung, *Nieuw. Arch. Wisk.* **15** (1927), 212-216.

P. Frankl
C.N.R.S.
Paris France

R.L. Graham
AT & T Bell Laboratories
Murray Hill,
New Jersey 07974

V. Rödl
Czech. Technical University
Husova, Czechoslovakia

7. Covering Complete Graphs by Monochromatic Paths

A. Gyárfás *

A theorem of R. Radó (see in [2]) says that if the edges of the countably infinite complete graph K_∞ are colored with r colors then the vertices of K_∞ can be covered by at most r (finite or one-way infinite) vertex-disjoint monochromatic paths. Edge-coloring theorems are usually extended from finite to infinite, however, in the case of Radó's theorem it seems natural to look for finite versions.

Conjecture 1. If the edges of a finite complete graph K are r-colored then the vertex-set of K can be covered by at most r vertex-disjoint monochromatic paths.

It is easy to see that Conjecture 1. is true for $r = 2$ (see in [1]) but for $r = 3$ it seems to be difficult. It is worth considering the following weaker versions.

Conjecture 2. If the edges of a finite complete graph K are r-colored then the vertex-set of K can be covered by at most r monochromatic paths.

Conjecture 3. There exists a function $f(r)$ with the following property: if the edges of a finite complete graph K are r-colored then the vertex-set of K can be covered by at most $f(r)$ vertex-disjoint monochromatic paths.

Conjectures 2 and 3 are both open even for $r = 3$. For general r we prove here the following result (which is weaker than Conjecture 2 or Conjecture 3).

Theorem. *There exists a function $f(r)$ with the following property: if the edges of a finite complete graph K are r-colored then the vertex-set of K can be covered by at most $f(r)$ monochromatic paths.*

The proof of this theorem is based on the following

* Supported by the AKA research Found of the Hungarian Academy of Sciences

Lemma. *Let r be a natural number and assume that a bipartite graph $G = (A, B)$ satisfies the following two conditions:*

(1) $rd(x) \geq |B|$ for all $x \in A$ ($d(x)$ denotes the degree of x)

(2) $$r\binom{r+1}{2}|A| \leq |B|$$

Then there exist at most r vertex-disjoint paths in G whose vertices cover A.

Proof. We may assume that $r \geq 2$. The required covering is found by the greedy algorithm. Let P_1 be a longest path of G starting in A and terminating in B. If $P_1, P_2, \ldots, P_{i-1}$ are already defined for some $i, 2 \leq i \leq r$, then P_i is defined as a longest path of G such that P_i and P_j are vertex-disjoint for all $j < i$, moreover P_i starts in A and terminates in B. The starting point of P_i is denoted by x_i and the number of vertices of P_i is denoted by $2k_i$. Note that $k_i \leq k_j$ if $i > j$, by definition.

Assume indirectly that $y \in A$ is not covered by any of the paths $P_1, P_2, \ldots, P_r$. The vertex x_i is not adjacent to any vertex of P_j for $j > i$, thus by (1) it is adjacent to all vertices of a set C_i, where $C_i \subset B, C_i \cap V(P_j) = \emptyset$ for all $j, 1 \leq j \leq r$, moreover

(3) $$|C_i| \geq r^{-1}|B| - \sum_{j=1}^{i} k_j \geq r^{-1}|B| - ik_1.$$

The sets C_i are pairwise disjoint by the choice of the paths P_i therefore the summing of (3) for $i = 1, 2, \ldots, r$ gives

(4) $$\left|\bigcup_{i=1}^{r} C_i\right| \geq |B| - \binom{r+1}{2}k_1.$$

The vertex y is not adjacent to any vertex of $\cup_{i=1}^{r} C_i$ by the definition of the paths P_i thus (1) implies

(5) $$r^{-1}|B| \leq d(y) \leq |B| - \left|\bigcup_{i=1}^{r} C_i\right|.$$

Comparing (4) and (5) we find that $r^{-1}\binom{r+1}{2}^{-1}|B| \leq k_1$ which contradicts (2) since $k_1 < |A|$ for $r \geq 2$. □

Proof of the theorem. Let X and Y be two disjoint subsets of the vertex-set of K such that $|X| = |Y| = \lfloor |K|/2 \rfloor = n$. Let G_i denote the bipartite subgraph of K induced by the edges of color i between X and Y. It is clear that we can define a partition of X into sets $X_1, X_2, \ldots, X_r$ such that $d_{G_i}(x) \geq r^{-1}n$ ($d_{G_i}(x)$ denotes the degree of x in G_i) holds for all $x \in X_i$. We can partition all X_i further in such a way that each part contains at most $r^{-1}\binom{r+1}{2}^{-1} n$ elements. If X_i' denotes one such part of X_i then the lemma can be applied to the bipartite graph (X_i', Y) induced by the edges of color i. Therefore X_i' can be covered by the vertices of at most r paths of color i. Applying this argument for each part of X_i and for all $i, 1 \leq i \leq r$, we get a covering of X by at most t monochromatic paths where

$$t = r\left(\sum_{i=1}^{r} \left\lceil |X_i| r \left(\binom{r+1}{2}\right) n^{-1} \right\rceil\right).$$

Since $\sum_{i=1}^{r} |X_i| = n$, we get $t \leq r^2(\binom{r+1}{2} + 1)$. Applying the same argument again with changing the role of X and Y, the theorem follows with

$$f(r) = 2r^2(\binom{r+1}{2} + 1) + 1).$$

□

References

[1] A.Gyárfás, Vertex Coverings by Monochromatic Paths and Cycles, *Journal of Graph Theory* **7** (1983),131-135.

[2] R.Rado, Monochromatic Paths in Graphs, *Annals of Discrete Math.* **3**, 191-194.

A.Gyárfás

Computer and Automation Institute

Hungarian Academy of Sciences

H-1111 Budapest, Kende u 13-17

8. Canonical Partition Behaviour of Cantor Spaces

Hanno Lefmann

Abstract

Using topological concepts A. Blass [1] proved an uncountable version of Ramsey's theorem. Here we prove a canonical version of this result.

1. Introduction and Statement of Results

In 1930 Ramsey [11] proved his famous pigeon-hole-principle for finite sets:

Theorem 1.1. [Ramsey 1930] *Let n, r be positive integers and let X be a countable infinite set. Then for every partition $\Delta : [X]^n \to \{0, 1, \ldots, r-1\}$ of the n–element subsets of X into r many classes there exists an infinite subset $Y \subseteq X$ such that the restriction $\Delta|[Y]^n$ is a constant mapping.*

A new situation arises if partitions into an arbitrary number of classes are considered. For this case Erdős and Rado [4] proved the so–called canonical version of Ramsey's theorem

Theorem 1.2. [Erdős, Rado 1950] *Let n be a positive integer and let X be a countable infinite set, which is totally ordered. Then for every mapping $\Delta : [X]^n \to X$ there exists an infinite subset $Y \subseteq X$ and a subset $I \subseteq \{0, \ldots, n-1\}$ such that for all $\{a_0, \ldots, a_{n-1}\}, \{b_0, \ldots, b_{n-1}\} \in [Y]^n$ with $a_0 < \ldots < a_{n-1}$ and $b_0 < \ldots < b_{n-1}$ it is valid:*

$$\Delta(\{a_0, \ldots, a_{n-1}\}) = \Delta(\{b_0, \ldots, b_{n-1}\})$$

iff

$$\{a_i | i \in I\} = \{b_i | i \in I\}.$$

□

For many other structures like for example arithmetic progressions, parameter–words and finite vector spaces, canonical partition results are known. A nice survey on this topic is given in [10].

In this paper we will consider uncountable versions of these theorems, especially for the underlying set 2^ω, the set of countable infinite 0,1–sequences. In [5] it was shown that there exist mappings $\Delta : [2^\omega]^2 \to \{0,1\}$ such that for every uncountable subset $Y \subseteq 2^\omega$ the restriction $\Delta|[Y]^2$ is not a constant mapping. But for these counterexamples the axiom of choice is used.

However, it turns out, if we consider mappings, which are in some sense constructive then a partition result will hold. This can be made precise by topological concepts, as was done by Blass in [1] by considering Baire–mappings $\Delta : [2^\omega]^n \to \{0,\ldots,r-1\}$.

Let 2^ω be endowed with the metric given by

$$d((\alpha_i)_{i<\omega},(\beta_i)_{i<\omega}) = \frac{1}{k+1} \quad \text{iff} \quad k = \min\{j < \omega \quad \alpha_j \neq \beta_j\}.$$

This yields the Cantor space with the usual Tychonoff product topology.

Let 2^ω be ordered lexicographically $\leq_{\text{lex}}$ with $0 < 1$. Finite subsets of 2^ω will usually be described by listing their elements in increasing order, where $\{\alpha_0,\ldots,\alpha_{n-1}\}_{<_{\text{lex}}}$ indicates $\alpha_0 <_{\text{lex}} \cdots <_{\text{lex}} \alpha_{n-1}$. With this description in mind, the set $[2^\omega]^n$ of n–element subsets of 2^ω can be identified with a subset of the product space $(2^\omega)^n$, from which it inherits a topology.

For subsets $\mathcal{T} \subseteq 2^\omega$ the set $[\mathcal{T}]^n$ is usually endowed with the topology induced by $[2^\omega]^n$.

A subset $\mathcal{T} \subseteq 2^\omega$ is perfect iff it is nonempty, closed and has no isolated points.

Let us visualize elements of 2^ω as infinite paths in the tree $2^{<\omega}$ (nodes being finite 0,1–sequences ordered by the initial segment relation). An infinite path in this tree is a totally ordered subset of $2^{<\omega}$ containing one node from each level. With this description in mind subsets $X \in [2^\omega]^n$ determine subtrees in $2^{<\omega}$. But two n–element subsets can have different shapes. The following picture shows how, typically, three–element subsets $\{\alpha_0,\alpha_1,\alpha_2\}_{<_{lex}} \subseteq 2^\omega$ look like.

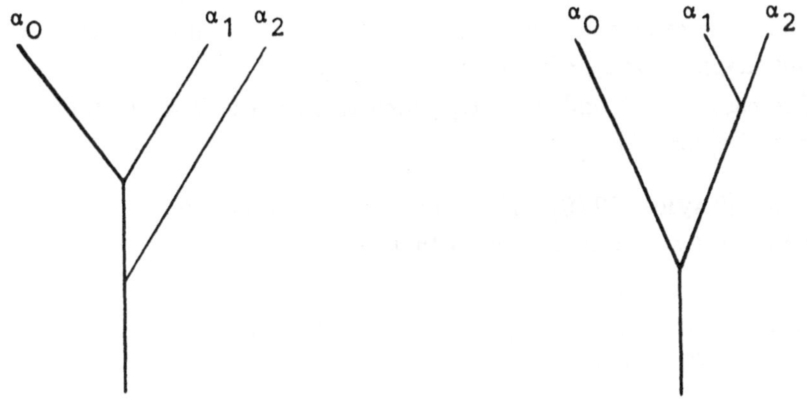

Figure 1

In the first case we have $d(\alpha_0, \alpha_1) < d(\alpha_1, \alpha_2)$, while in the second $d(\alpha_0, \alpha_1) > d(\alpha_1, \alpha_2)$. Thus for perfect sets partition results can only be expected w.r.t. subsets of fixed shape. To be more concrete we need the following

Notation. Let n be a positive integer and let $(\{1, \ldots, n-1\}, \preceq)$ be a total order. Further let $X \subseteq 2^\omega$. Then

$$[X]^n_\preceq = \{\{\alpha_0, \ldots, \alpha_{n-1}\}_{<_{lex}} \in [X]^n \mid d(\alpha_{i-1}, \alpha_i) < d(\alpha_{j-1}, \alpha_j) \text{ iff } i \prec j \text{ for all } 1 \leq i < j \leq n-1\}$$

where $\leq$ denotes the usual order on the set $\mathbf{R}$ of real numbers.

For the special case $n = 3$ and $1 \prec 2$ we have $[X]^3_\prec = \{\{\alpha_0, \alpha_1, \alpha_2\}_{<_{lex}} \in [X]^3 | d(\alpha_0, \alpha_1) < d(\alpha_1, \alpha_2)\}$. In general, there are $(n-1)!$ total orders on the set $\{1, \ldots, n-1\}$. Thus there are $(n-1)!$ such subsets of $[X]^n$, determined by the relative distances of consecutive elements.

Galvin [6] discovered the following partition result for $n \leq 3$, compare also [2]. The general case for arbitrary n was settled by Blass [1].

Theorem 1.3. [**Blass 1981**]. *Let n, r be positive integers and let $\mathcal{T} \subseteq 2^\omega$ be a perfect subset. Then for every Baire–mapping $\Delta : [\mathcal{T}]^n \to \{0, \ldots, r-1\}$ there exists a perfect subset $\mathcal{P} \subseteq \mathcal{T}$ such that $\Delta | [\mathcal{P}]^n_\preceq =$ const. for every total order $(\{1, \ldots, n-1\}, \preceq)$.*

Thus images of n–element subsets $A \subset \mathcal{P}$ only depend on their order–type, which is the total order on $\{1, \ldots, n-1\}$ induced by the relative distances of consecutive elements of A.

In the following we investigate Baire–mappings $\Delta : [2^\omega]^n \to X$, where X is an arbitrary metric space. Consider first the case $n = 1$. For simplicity, let $\Delta : 2^\omega \to X$ be a continuous mapping. This induces another continuous mapping $\Delta^* : [2^\omega]^2 \to \{0,1\}$ by $\Delta(\{\alpha_0, \alpha_1\}) = 0$ iff $\Delta(\alpha_0) = \Delta(\alpha_1)$. Theorem 1.3 implies the existence of a perfect subset $\mathcal{P} \subseteq 2^\omega$ such that either $\Delta|\mathcal{P}$ is a constant mapping or $\Delta|\mathcal{P}$ is an injective mapping.

The case $n = 2$ and $X = \omega$, endowed with the discrete topology, was settled by Taylor [12]:

Theorem [Taylor 1979]. *For every Baire–mapping $\Delta : [2^\omega]^2 \to \omega$ there exists a perfect subset $\mathcal{P} \subseteq 2^\omega$ such that*

either $\Delta|[\mathcal{P}]^2 = \text{const}$

or $\Delta(\{\alpha_0, \alpha_1\}_{<_{lex}}) = \Delta(\{\beta_0, \beta_1\}_{<_{lex}})$ iff $d(\{\alpha_0, \alpha_1\}) = d(\{\beta_0, \beta_1\})$ for all $\{\alpha_0, \alpha_1\}, \{\beta_0, \beta_1\} \in [\mathcal{P}]^2$. □

If we consider Baire–mappings $\Delta : [2^\omega]^2 \to X$, where X is an arbitrary metric space, additional patterns occur. This can be seen via the continuous mappings $\Delta_I : [2^\omega]^2 \to [2^\omega]^{|I|}$ and $\Delta_j^d : [2^\omega]^2 \to 2^\omega \times \mathbf{R}$, where $\emptyset \neq I \subseteq \{0,1\}$ and $j = 0$ or $j = 1$, defined by

$$\Delta_I(\{\alpha_0, \alpha_1\}_{<_{lex}}) = \{\alpha_i | i \in I\}$$

and

$$\Delta_j^d(\{\alpha_0, \alpha_1\}_{<_{lex}}) = (\alpha_j, d(\alpha_0, \alpha_1)).$$

The corresponding patterns can not occur in Taylor's result, since he considered the case $X = \omega$. For arbitrary metric spaces X a canonical partition result for Baire–mappings $\Delta : [2^\omega]^2 \to X$ has to respect beside the two patterns in Taylor's theorem at least the five patterns arising from the mappings Δ_I and Δ_j^d. The following result shows that these seven patterns are sufficient to describe the canonical partition behaviour for the case $n = 2$:

Theorem. *Let X be a metric space and let $\Delta : [2^\omega]^2 \to X$ be a Baire–mapping. Then there exists a perfect subset $\mathcal{P} \subseteq 2^\omega$ and subsets $I \subseteq \{0,1\}$ and $J \subseteq \{1\}$ with $J = \emptyset$ for $I = \{0,1\}$ such that for all $\{\alpha_0, \alpha_1\}_{<_{lex}}, \{\beta_0, \beta_1\}_{<_{lex}} \in [2^\omega]^2$ it is valid:*

$$\Delta(\{\alpha_0, \alpha_1\}_{<_{lex}}) = \Delta(\{\alpha_0, \alpha_1\}_{<_{lex}})$$

iff $\{\alpha_i | i \in I\} = \{\beta_i | i \in I\}$ and $\{d(\alpha_{j-1}, \alpha_j) | j \in J\} = \{d(\beta_{j-1}, \beta_j) | j \in J\}$. □

The condition "$I = \{0,1\}$ implies $J = \emptyset$" is used to avoid redundancies. Obviously, for $X = \omega$ we get Taylor's result.

For the general case of partitioning arbitrary n–element subsets A of 2^ω it turns out that the canonical patterns are determined by subsets of A and subsets of the set of distances between consecutive elements of A. As in theorem 1.3 we can start with a ground set $\mathcal{T}$, which is a perfect subset of 2^ω.

Theorem 1.4. *Let n be positive integer and let $(\{1,\ldots,n-1\},\preceq)$ be a total order. Further let $\mathcal{T} \subseteq 2^\omega$ be a perfect subset and let X be a metric space. Then for every Baire–mapping $\Delta : [\mathcal{T}]^n_\prec \to X$ there exists a perfect subset $\mathcal{P} \subseteq \mathcal{T}$ and subsets $I \subseteq \{0,\ldots,n-1\}$ and $J \subseteq \{1,\ldots,n-1\}$ such that for every $A,B \in [\mathcal{P}]^n_\prec$ with $A = \{\alpha_0,\ldots,\alpha_{n-1}\}_{<_{lex}}$ and $B = \{\beta_0,\ldots,\beta_{n-1}\}_{<_{lex}}$ it is valid:*

$$\Delta(A) = \Delta(B)$$

iff

$$\{\alpha_i | i \in I\} = \{\beta_i | i \in I\} \text{ and}$$

$$\{d(\alpha_{j-1},\alpha_j) | j \in J\} = \{d(\beta_{j-1},\beta_j) | j \in J\}.$$

□

So far we have considered Baire–mappings $\Delta : [\mathcal{T}]^n \to X$ w.r.t. subsets of one order–type only. Nothing is said about the behaviour of such mappings w.r.t. different types. In [13] general concepts for this situation have been developed. Voigt showed that categories satisfying certain conditions have the so–called diversification property; roughly speaking this means that we can achieve that two mappings either have disjoint images or two objects have the same image iff their patterns are the same. By applying the methods developed in [13, especially Lemma 3] to our situation we get immediately:

Theorem 1.5. *Let n be a positive integer, let $\mathcal{T} \subseteq 2^\omega$ be perfect and let X be a metric space. Then for every Baire–mapping $\Delta : [\mathcal{T}]^n \to X$ there exists a perfect subset $\mathcal{P} \subseteq \mathcal{T}$ and for every total order $(\{1,\ldots,n-1\},\preceq)$ there exist subsets $I_\prec \subseteq \{0,\ldots,n-1\}$ and $J_\prec \subseteq \{1,\ldots,n-1\}$ such that for every pair $(\preceq^*,\preceq^{**})$ of total orders on $\{1,\ldots,n-1\}$ one of the following two possibilities holds:*

(i) $\Delta(A) \neq \Delta(B)$ *for all* $A \in [\mathcal{P}]^n_{\prec^*}, B \in [\mathcal{P}]^n_{\prec^{**}}$ *or*

(ii) $\Delta(\{\alpha_0,\ldots,\alpha_{n-1}\}_{<_{lex}}) = \Delta(\{\beta_0,\ldots,\beta_{n-1}\}_{<_{lex}})$ *iff* $\{\alpha_i | i \in I_{\prec^*}\} = \{\beta_i | i \in I_{\prec^{**}}\}$ and $\{d(\alpha_{j-1},\alpha_j) | j \in J_{\prec^*}\} = \{d(\beta_{j-1},\beta_j) | j \in J_{\prec^{**}}\}$ *for all* $\{\alpha_0,\ldots,\alpha_{n-1}\}_{<_{lex}} \in [\mathcal{P}]^n_{\prec^*}, \{\beta_0,\ldots,\beta_{n-1}\}_{<_{lex}} \in [\mathcal{P}]^n_{\prec^{**}}$. □

Notice that for identical total orders $\prec^*$ and $\prec^{**}$ necessarily the second case occurs.

2. Proof of Theorem 1.4

Let $2^{<\omega}$ be the set of finite 0, 1–sequences. For sequences $\alpha, \beta \in 2^{<\omega}$ with $\alpha = (\alpha_0, \ldots, \alpha_{m-1})$ and $\beta = (\beta_0, \ldots, \beta_{n-1})$ let $\alpha \otimes \beta = (\alpha_0, \ldots, \alpha_{m-1}, \beta_0, \ldots, \beta_{n-1})$ be the concatenation of α and β.

For different infinite sequences $\gamma, \delta \in 2^\omega$ let $\gamma \cap \delta$ be the unique finite sequence $\alpha \in 2^{<\omega}$, such that α is an initial segment of both γ and δ and there is no longer sequence with this property.

The description of the possible patterns in Theorems 1.4 and 1.5 is redundant. This can be avoided, if we allow only canonical pairs (I, J) of subsets.

Definition 2.1. Let $\mathcal{T} \subseteq 2^\omega$ be a perfect subset. For a positive integer n let $(\{1, \ldots, n-1\}, \preceq)$ be a total order. Further let $I \subseteq \{0, \ldots, n-1\}$ and $J \subseteq \{1, \ldots, n-1\}$ be subsets. The pair (I, J) is a *canonical pair* (w.r.t. $(\{1, \ldots, n-1\}, \preceq)$) iff for every $A \in [\mathcal{T}]^n_\preceq$ with $A = \{\alpha_0, \ldots, \alpha_{n-1}\}_{<_{lex}}$ it is valid:

(i) $d(\alpha_k, \alpha_l) \in \{d(\alpha_{j-1}, \alpha_j) | j \in J\}$ for every $k, l \in I$ with $k \neq l$.

(ii) $\{\alpha_{j-1} \cap \alpha_j | j \in J\}$ is closed w.r.t. taking infima; i.e. if $j, j^* \in J$ with $j \neq j^*$ then for the unique $k < n$ with $d(\alpha_{k-1}, \alpha_k) = d(\alpha_j, \alpha_{j^*})$ we have: $k \in J$.

It will be convenient to work with skew subsets $\mathcal{T} \subseteq 2^\omega$, where $\mathcal{T}$ is *skew* iff for every $\alpha_0, \alpha_1, \beta_0, \beta_1 \in 2^\omega$ with $\alpha_0 \neq \alpha_1$ and $\beta_0 \neq \beta_1$ it is valid $d(\alpha_0, \alpha_1) = d(\beta_0, \beta_1)$ implies $\alpha_0 \cap \alpha_1 = \beta_0 \cap \beta_1$.

By straightforward arguments the following can be shown (for a proof compare [1]):

Lemma 2.2. *Every perfect subset of 2^ω contains a skew perfect subset.*

Notation. Let $A \subseteq 2^\omega$ be finite with $A = \{\alpha_0, \ldots, \alpha_{n-1}\}_{<_{lex}}$. Let $D(A) = \{d(\alpha_{i-1}, \alpha_i) | 1 \leq i < n\}$ be the set of distances between consecutive elements of A. Further let $I \subseteq \{0, \ldots, n-1\}$ and $J \subseteq \{1, \ldots, n-1\}$ be subsets. Then

$$A : I = \{\alpha_i | i \in I\}$$

is the I–subset of A and

$$D(A) : J = \{d(\alpha_{j-1}, \alpha_j) | j \in J\}$$

is the J–subset of $D(A)$.

Now we can give an irredundant description of the canonical cases of Blass' theorem:

Theorem 1.4'. *Let n be a positive integer and let $(\{1, \ldots, n-1\}, \preceq)$ be a total order. Further let $\mathcal{T} \subseteq 2^\omega$ be a perfect subset and let X be a metric space. Then for every Baire–mapping $\Delta : [\mathcal{T}]^n_\preceq \to X$ there exist a perfect skew* *subset* $\mathcal{P} \subseteq \mathcal{T}$ *and a canonical pair* (I, J) *of subsets, where* $I \subseteq \{0, \ldots, n-1\}$ *and* $J \subseteq \{1, \ldots, n-1\}$*, such that for every* $A, B \in [\mathcal{P}]^n_{\preceq}$ *it is valid:*

$$\Delta(A) = \Delta(B)$$

iff

$$A : I = B : I \text{ and } D(A) : J = D(B) : J.$$

In the same way theorem 1.5 can be adapted.

Before we prove theorem 1.4' we will show that it suffices to consider continuous mappings $\Delta : [\mathcal{T}]^n_{\preceq} \to X$. Kuratowsky proved that for metric spaces X, Y, where X is separable, a mapping $\Delta : Y \to X$ is a Baire–mapping iff there exists a meager set $M \subseteq Y$ such that the restriction $\Delta | Y \setminus M$ is continuous (compare [7]). By a result in [3] we can drop in the case $Y = [\mathcal{T}]^n_{\preceq}$ the assumption that X is separable. In [8] it was shown, that for any meager set M in $[X]^n$, where X is a complete metric space without isolated points, there is a perfect subset $P \subseteq X$ such that $[P]^n \cap M = \emptyset$. Thus we have

Lemma 2.3. *Let* n *be a positive integer and let* $(\{1, \ldots, n-1\}, \preceq)$ *be a total order. Further let* $\mathcal{T} \subseteq 2^\omega$ *be perfect and skew and let* X *be a metric space. Then for every Baire–mapping* $\Delta : [\mathcal{T}]^n_{\preceq} \to X$ *there exists a perfect skew subset* $\mathcal{P} \subseteq \mathcal{T}$ *such that* $\Delta | [\mathcal{P}]^n_{\preceq}$ *is continuous.*

Proof of theorem 1.4'. By lemma 2.3 we can restrict to continuous mappings $\Delta : [T]^n \to X$. We proceed by induction on n. Although the case $n = 1$ can be derived from a stronger result of Silver (compare also [9]) we show it to give an idea for the proof in the general case.

Let $\Delta : \mathcal{T} \to X$ be a continuous mapping. This induces another mapping $\Delta^* : [\mathcal{T}]^2 \to \{0, 1\}$ by $\Delta^*(\{\alpha_0, \alpha_1\}) = 0$ iff $\Delta(\alpha_0) = \Delta(\alpha_1)$. Clearly, Δ^* is a continuous mapping. By theorem 1.3 there exist a perfect skew subset $\mathcal{P} \subseteq \mathcal{T}$ and an $i < 2$ such that $\Delta^*([\mathcal{P}]^2) = i$. Since each two different elements of $\mathcal{P}$ determine an element of $[\mathcal{P}]^2$ we have

— $\Delta | \mathcal{P}$ =const. for $i = 0$ and

— $\Delta | \mathcal{P}$ is one–to–one for $i = 1$.

For establishing the inductive step we use two special constructions. Let $T \in [\mathcal{T}]^n_{\preceq}$ with $T = \{\alpha_0, \ldots, \alpha_{n-1}\}_{<_{lex}}$. Define a mapping $f : \{0, \ldots, n-1\} \to \{0, \ldots, n-1\}$ where $f(i)$ is the unique $i + j$ with $j \in \{-1, 1\}$ such that $\alpha_i \cap \alpha_{i-j}$ is an initial segment of $\alpha_i \cap \alpha_{i+j}$.

Example.

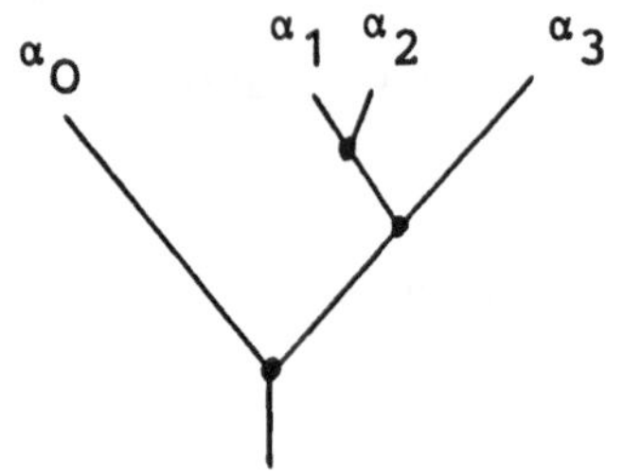

Figure 2

Here we have $f(0) = 1, f(1) = 2, f(2) = 1, f(3) = 2$.

Definition of $T + i$. For $i < n$ let $T + i \in [\mathcal{T}]^{n+1}$ result from T by adding a new element $\alpha_i^+ \in \mathcal{T}$ (in some sense a copy of α_i) to T such that the following is valid:

(i) $\alpha_i^+ <_{lex} \alpha_i$

(ii) $\alpha_i \cap \alpha_i^+$ is an initial segment of $\alpha_i \cap \alpha_{f(i)}$

(iii) for $j = 0, 1$ it holds:
if $(\alpha_i \cap \alpha_{f(i)}) \otimes j$ is an initial segment of α_i, then $(\alpha_i^+ \cap \alpha_{f(i)}) \otimes j$ is an initial segment of α_i^+; thus α_i^+ leaves $\alpha_{f(i)}$ in the same direction as α_i does.

(iv) T and $T \setminus \{\alpha_i\} \cup \{\alpha_i^+\}$ have the same order–type.

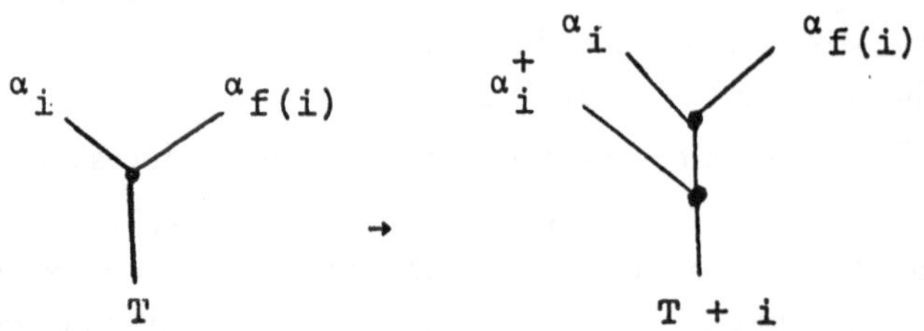

Figure 3

Definition of $T \oplus i$. For $i < n$ let $T \oplus i \in [\mathcal{T}]^{n+1}$ result from T by adding a new element $\alpha_i^\oplus \in \mathcal{T}$ (in some sense a copy of α_i) in the following way:

(i) $\alpha_i^\oplus <_{lex} \alpha_i$

(ii) $\alpha_i \cap \alpha_{f(i)}$ is a proper initial segment of $\alpha_i \cap \alpha_i^\oplus$

(iii) T and $T \setminus \{\alpha_i\} \cup \{\alpha_i^{\oplus}\}$ have the same order–type.

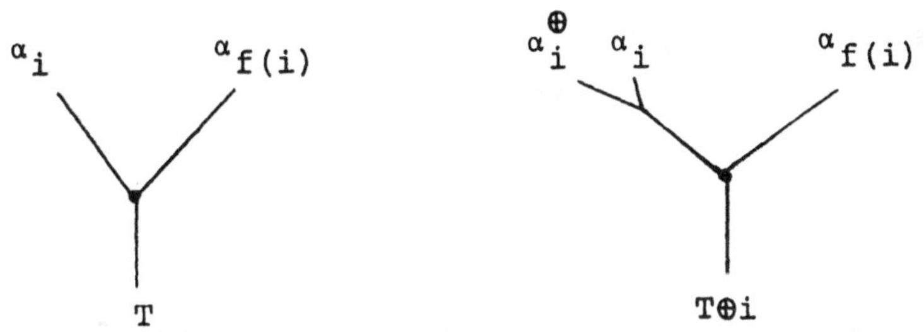

Figure 4

Let the resulting sets $T+i$ resp. $T \oplus i$ have order–types $(\{1,\ldots,n\}; \preceq_i^+)$ resp. $(\{1,\ldots,n\}; \preceq_i^{\oplus})$.

$T+i$ or $T \oplus i$ need not exist for every $T \in [\mathcal{T}]^n_{\preceq}$, but certainly there exist subsets $T \in [\mathcal{T}]^n_{\preceq}$ such that $T+i$ and $T \oplus i$ exist. If they exist, they are in general not unique, while their order–types, and only these will be of interest in the following, are independent of the special choice of the added elements.

Now we start with the inductive step. Let $n \geq 2$ be given and assume that the theorem is valid for all $k < n$. Let $\Delta : [\mathcal{T}]^n_{\preceq} \to X$ be a continuous mapping. For every $i < n$ the mapping Δ induces another mapping $\Delta_i^+ : [\mathcal{T}]^{n+1}_{\preceq_i^+} \to \{0,1\}$ by $\Delta_i^+(T) = 0$ iff $\Delta(T : (\{0,\ldots,n\} \setminus \{i\})) = \Delta(T : (\{0,\ldots,n\} \setminus \{i+1\}))$. Since the Δ_i^+ are continuous mappings, by theorem 1.3 there exists a perfect skew subset $\mathcal{T}_0 \subseteq \mathcal{T}$ and for every $i < n$ there exist numbers $c_i < 2$ such that $\Delta_i^+([\mathcal{T}]^{n+1}_{\preceq_i^+}) = c_i$.

We distinguish two cases:

Case 1. $c_i = 0$ for some $i < n$.

For $T \in [\mathcal{T}_0]^n_{\preceq}$ let the resulting set $T : (\{0,\ldots,n-1\} \setminus \{i\})$ have order–type $(\{1,\ldots,n-2\}, \prec^*)$. Then Δ induces a mapping $\Delta^* : [\mathcal{T}_0]^{n-1}_{\prec^*} \to X$ by $\Delta^*(T : (\{0,\ldots,n-1\} \setminus \{i\})) = \Delta(T)$. By eventually restricting to a perfect subset of $\mathcal{T}_0$ we can assume that Δ^* is well defined and continuous.

By induction hypothesis there exists a perfect subset $\mathcal{T}_1 \subseteq \mathcal{T}_0$ and subsets $I^* \subseteq \{0,\ldots,n-2\}$ and $J^* \subseteq \{1,\ldots,n-2\}$ such that for all $T_0, T_1 \in [\mathcal{T}_1]^{n-1}_{\prec^*}$ it is valid:

$$\Delta^*(T_0) = \Delta^*(T_1)$$

iff

$$T_0 : I^* = T_1 : I^* \text{ and } D(T_0) : J^* = D(T_1) : J^*.$$

Now let $I \subseteq \{0,\ldots,n-1\}$ and $J \subseteq \{1,\ldots,n-1\}$ be such that we have for $T \in [\mathcal{T}_1]^n_{\preceq}$:

$$T : (\{0,\ldots,n-1\} \setminus \{i\}) : I^* = T : I$$

and

$$D(T:(\{0,\dots,n-1\}\setminus\{i\})):J^* = D(T):J$$

It follows for $T_0, T_1 \in [\mathcal{T}_1]^n_{\prec}$:

$$\Delta(T_0) = \Delta(T_1)$$

iff

$$T_0 : I = T_1 : I \text{ and } D(T_0):J = D(T_1):J.$$

Case 2. $c_i = 1$ for every $i < n$.

For every $i < n$ the mapping Δ induces a mapping $\Delta_i^{\oplus} : [\mathcal{T}_0]^{n+1}_{\prec^{\oplus}} \to \{0,1\}$ by

$$\Delta_i^{\oplus}(T) = 0$$

iff $\Delta(T:(\{0,\dots,n\}\setminus\{i\})) = \Delta(T:(\{0,\dots,n\}\setminus\{i+1\}))$.

Each Δ_i is a continuous mapping. By theorem 1.3 there exist a perfect skew subset $\mathcal{T}_1 \subseteq \mathcal{T}_0$ and for every $i < n$ numbers $d_i < 2$ such that $\Delta_i^{\oplus}([\mathcal{T}_1]^{n+1}_{\prec_i^{\oplus}}) = d_i$.

Consider for every two–element set $\{T_0, T_1\} \in [[\mathcal{T}_1]^n_{\prec}]^2$ the union $T_0 \cup T_1$, say $|T_0 \cup T_1| = p$, of order–type $(\{1,\dots,p-1\};\preceq)$ and let $I_0, I_1 \subseteq \{0,\dots,p-1\}$ be subsets such that $(T_0 \cup T_1) : I_0 = T_0$ and $(T_0 \cup T_1) : I_1 = T_1$. Let $((\{1,\dots,p_i\};\preceq_i), I_i^0, I_i^1)_{i<q}$ be an injective enumeration of the occurring triples. For every $i < q$ let $\Delta_i : [\mathcal{T}_1]^{p_i}_{\prec} \to 2$ be mappings defined by $\Delta_i(T) = 0$ iff $\Delta(T : I_i^0) = \Delta(T : I_i^1)$.

Since the Δ_i are continuous mappings, we get by theorem 1.3 a perfect skew subset $\mathcal{T}_2 \subseteq \mathcal{T}_1$ such that $\Delta_i | [\mathcal{T}_2]^{p_i}_{\prec_i}$ =const. for every $i < q$. Thus the patterns w.r.t. Δ of every two subsets $T_0, T_1 \in [\mathcal{T}_2]^n_{\prec}$, which have the same relative position to each other, are the same.

Let $I = \{i < n | d_i = 1\}$. We claim that for all $T_0, T_1 \in [\mathcal{T}_2]^n_{\prec}$ it is valid:

$$\Delta(T_0) = \Delta(T_1)$$

iff

$$T_0 : I = T_1 : I \text{ and } D(T_0) = D(T_1).$$

For $i = 0,1$ let $T_i = \{\alpha_0^i,\dots,\alpha_{n-1}^i\}_{<_{lex}} \in [\mathcal{T}_2]^n_{\prec}$ be given. At first we show the implication from right to left. So let $T_0 : I = T_1 : I$ and $D(T_0) = D(T_1)$. If $I = \{0,\dots,n-1\}$, there is nothing to prove. So let $I \neq \{0,\dots,n-1\}$ and take $l \in \{0,\dots,n-1\}\setminus I$. Since $d_l = 0$, we conclude

$$\Delta(\{\alpha_0^0,\dots,\alpha_{n-1}^0\}_{<_{lex}}) = \Delta(\{\alpha_0^0,\dots,\alpha_{n-1}^0\}\setminus\{\alpha_l^0\}\cup\{\alpha_l^1\}).$$

Iterating this we get

$$\Delta(\{\alpha_0^0,\dots,\alpha_{n-1}^0\}_{<_{lex}}) = \Delta(\{\alpha_i^0 | i \in I\} \cup \{\alpha_j^1 | j \in \{0,\dots,n-1\}\setminus I\})$$

and thus $\Delta(T_0) = \Delta(T_1)$.

Now we prove the implication from left to right. Let $\Delta(T_0) = \Delta(T_1)$ and assume $T_0 : I \neq T_1 : I$ or $D(T_0) \neq D(T_1)$. Suppose $D(T_0) \neq D(T_1)$. Let $d = \min((D(T_0) \setminus D(T_1)) \cup (D(T_1) \setminus D(T_0)))$, where w.l.o.g. $d = d(\alpha^0_{i-1}, \alpha^0_i)$.

Since d is minimal, there is $j < 2$ such that $\alpha = (\alpha^0_{i-1} \cap \alpha^0_i) \otimes j$ is not an initial segment for any $\beta \in T_1$. Let α be an initial segment of, say, α^0_i. Choose $\hat{\alpha}_i \in \mathcal{T}_2 \setminus (T_0 \cup T_1)$ such that

— $(\alpha^0_{i-1} \cap \hat{\alpha}_i) \otimes j$ is an initial segment of $\hat{\alpha}_i$

— α is not an initial segment of $\hat{\alpha}_i$

— $T_0 \cup T_1$ and $(T_0 \cup T_1) \setminus \{\alpha^0_i\} \cup \{\hat{\alpha}_i\}$ have the same order–type.

By eventually choosing other sets T_0, T_1 this is possible, since all pairs of subsets with the same relative position have the same pattern w.r.t. Δ. But then we have

$$\Delta(T_0) = \Delta(T_1) = \Delta(T_0 \setminus \{\alpha^0_i\} \cup \{\alpha^1_i\})$$

contradicting $c_i = 1$ for $i < n$.

Suppose now $D(T_0) = D(T_1)$ but $T_0 : I \neq T_1 : I$. Then there exists $i \in I$ with $\alpha^0_i \neq \alpha^1_i$, say $\alpha^0_i <_{lex} \alpha^1_i$. Choose $\alpha \in \mathcal{T}_2 \setminus (T_0 \cup T_1)$ with $\alpha^0_i <_{lex} \alpha <_{lex} \alpha^1_i$ such that $T_0 \cup T_1$ and $(T_0 \cup T_1) \setminus \{\alpha^1_i\} \cup \{\alpha\}$ have the same order type. As before we get

$$\Delta(T_0) = \Delta(T_1) = \Delta(T_1 \setminus \{\alpha^1_i\} \cup \{\alpha\}),$$

which contradicts $d_i = 1$. This finishes the proof of theorem 1.4. □

3.Applications

Let $\mathbf{R}$ denote the set of real numbers, endowed with the usual euclidean metric. By using binary expansion of real numbers, i.e. using the continuous mapping $f : 2^\omega \to [0,1]$ with

$$f((a_i)_{i<\omega}) = \sum_{i<\omega} a_i 2^{-i-1}$$

we conclude from theorem 1.5:

Corollary 3.1. *Let n be a positive integer. Let $\mathcal{T} \subseteq \mathbf{R}$ be a perfect subset and let X be a metric space. Then for every Baire–mapping $\Delta : [\mathcal{T}]^n \to X$ there exists a perfect subset $\mathcal{P} \subseteq \mathcal{T}$ and for every total order $(\{1, \ldots, n-1\}; \preceq)$ there exist subsets $I_\prec \subseteq \{0, \ldots, n-1\}$ and $J_\prec \subseteq \{1, \ldots, n-1\}$ such that for every pair $(\preceq^*, \preceq^{**})$ of total orders on $\{1, \ldots, n-1\}$ one of the following two possibilities holds:*

(i) $\Delta(A) \neq \Delta(B)$ *for all* $A \in [\mathcal{P}]^n_{\prec^*}$ *and* $B \in [\mathcal{P}]^n_{\prec^{**}}$

(ii) $\Delta(A) = \Delta(B)$ iff $A : I_{\prec^\bullet} = B : I_{\prec^{\bullet\bullet}}$ and

$$\{k \in \mathbf{Z} | 2^{k-1} \leq a_j - a_{j-1} < 2^k \text{ for some } j \in J_{\prec^\bullet}\} = $$
$$= \{k \in \mathbf{Z} | 2^{k-1} \leq b_j - b_{j-1} < 2^k \text{ for some } j \in J_{\prec^{\bullet\bullet}}\}$$

for all $A \in [P]^n_{\prec^\bullet}$ and $B \in [P]^n_{\prec^{\bullet\bullet}}$ *with* $A = \{a_0, \ldots, a_{n-1}\}_<$ and $B = \{b_0, \ldots, b_{n-1}\}_<$. □

Thus the canonical patterns are determined by subsets of the partitioned sets and by subsets of the set of distances between consecutive elements.

References

[1] A. Blass, A partition theorem for perfect sets, *Proc. Amer. Math. Soc.*, **82** (1981) 271-277.

[2] J.P. Burgess, A selector principle for $\sum_1^1$ equivalence relations, *Michigan Math. J.* **24** (1977) 65-76.

[3] A. Emerik, R. Frankiewicz and W. Kulpa, On functions having the Baire–property, *Bull. Acad. Polon. Math.*, **27** (1979) 489-491.

[4] P. Erdős and R. Rado, A combinatorial theorem, *J. London Math. Soc.* **25** (1950) 249-255.

[5] P. Erdős and R. Rado, Combinatorial theorems on classification of subsets of a given set, *Proc. London Math. Soc.* **3** (1952) 417-439.

[6] F. Galvin, Partition theorems for the real line, *Notices Amer. Math. Soc.* **15** (1968) 660; Errata: *Notices Amer. Math. Soc.* **16** (1969) 1095.

[7] K. Kuratowski, *Topology I*, Academic Press, 1966, New York.

[8] J. Mychielsky, Independent sets in topological algebras, *Fund. Math.* **55** (1964) 139-147.

[9] H.J. Prömel, S.G. Simpson and B. Voigt, A dual form of Erdős–Rado's canonizing theorem, *J. Comb. Theory Ser. A* **42**, (1986) 159-178.

[10] H.J. Prömel and B. Voigt, Canonizing Ramsey Theory, *preprint*, Bielefeld, 1985.

[11] F.P. Ramsey, On a problem of formal logic, *Proc. London Math. Soc.* **30** (1930), 264-286.

[12] A.D. Taylor, Partitions of pairs of reals, *Fund. Math.* **99** (1979) 51-59.

[13] B. Voigt, Canonizing partition theorems: diversifications, products and iterated versions, *J. Comb. Th. Ser. A* **40** (1985) 349-376.

H. Lefmann
Universität Bielefeld
Bielefeld

9. Extremal Problems for Discrepancy

L. Lovász and K. Vesztergombi

Abstract

We determine the maximum number of edges in a hypergraph with a given number of vertices and given hereditary discrepancy. We derive bounds on the maximum number of rows in an integral matrix with a given number of columns and with given hereditary discrepancy. The main tool is a geometric interpretation of discrepancy and a relation between the volumes of polar convex bodies.

1. A Geometric Formulation of Discrepancy

A matrix A is called *totally unimodular* if every non-singular square submatrix of it has determinant ± 1. Heller (1957) showed that a totally unimodular matrix A of n columns can have at most $1 + (n+1)n$ distinct rows. If all entries of a matrix A are 0's and 1's then it can be considered as the vertex-edge incidence-matrix of a hypergraph H. We call a hypergraph *unimodular* if its incidence-matrix is totally unimodular. Heller's theorem implies that a unimodular hypergraph on n vertices has at most $\binom{n}{2} + n + 1$ edges. The maximum is attained by the hypergraph consisting of all intervals of an n-element ordered set.

The *discrepancy* of a hypergraph H is defined by Beck (1981) as the smallest integer d for which the vertices of H can be 2-colored with red and blue so that the difference between the number of red and blue vertices in each edge of H is at most d. It follows from elementary properties of total unimodularity that every unimodular hypergraph has discrepancy ≤ 1. Conversely, a theorem of Ghouila-Houri (1962) implies that if every restriction of a hypergraph has discrepancy ≤ 1 then the hypergraph is totally unimodular. (The *restriction* of H to a subset W of the vertex set S of H is the hypergraph $H_W = \{W \cap X : X \in H\}$.) So, a natural way to generalize unimodularity is to introduce hereditary discrepancy, which is obtained by considering the discrepancy not only the 2-colorations of the set S, but also of the restrictions to subsets of S. For technical reasons which will become apparent later, we divide this number by 2. Exactly, we define the ***discrepancy*** of H by

$$\mathrm{disc}(H) = \min_{T\subset S}\max_{E\in H} \left||T\cap E| - \frac{1}{2}|E|\right|$$

and the ***hereditary discrepancy*** by

$$\mathrm{herdisc}(H) = \max_{W\subset S} \mathrm{disc}(H_W)$$

In Lovász, Spencer and Vesztergombi (1986) the linear discrepancy of a hypergraph is also defined:

$$\mathrm{lindisc}(H) = \max_{c\in[0,1]^S}\min_{T\subset S}\max_{E\in H} \left||T\cap E| - c(E)\right|$$

where $c(E) = \sum_{v\in E} c_v$.

These notions can be extended to general matrices A (replacing the incidence matrix of H). For any $m\times n$ matrix A, we define

$$\mathrm{disc}(A) = \min_{x\in\{0,1\}^n} \|A(x - \frac{1}{2}\mathbf{1})\|_\infty .$$

where $\mathbf{1}$ is the vector with all entries 1. We also define

$$\mathrm{herdisc}(A) = \max_{A'} \mathrm{disc}(A'),$$

where A' ranges through all submatrices of A. Since deleting rows from the matrix does not increase the discrepancy, it suffices to consider here submatrices arising by deleting columns. For incidence matrices of hypergraphs, this corresponds to the restriction of the hypergraph.

The generalization of the linear discrepancy to matrices is also quite natural:

$$\mathrm{lindisc}(A) = \max_{c\in\{0,1\}^1}\min_{x\in\{0,1\}^n} \|A(x-c)\|_\infty .$$

Similarly as herdisc, *herlindisc* can be defined.

Let K be a centrally symmetric convex set in $\mathbf{R}^n$, centered at 0, and let $t > 0$. Consider the body tK and all bodies obtained by translating the center of tK to the vertices of the unit cube $[0,1]^n$. Let disc(K) be the smallest t for which the union of these translates covers the center of the unit cube. We define herdisc(K) as the maximum of disc$(K\cap\Sigma)$, where Σ ranges through all coordinate subspaces. Another way to say this is that herdisc(K) is the smallest t for which the center of each face of the unit cube is contained in one of the copies of tK translated to the vertices of the face. Furthermore, we define lindisc(K) as the least t for which the translates of tK to the vertices of the unit cube cover the whole cube.

These geometric discrepancy notions generalize the discrepancies of a matrix A if we apply them to the convex body $K = U_A = \{x\in\mathbf{R}^n : \|Ax\|_\infty \le 1\}$.

In (Lovász, Spencer and Vesztergombi, 1986) it was proved that for any convex body K,

(1) $$\mathrm{lindisc}(K) \leq 2\mathrm{herdisc}(K).$$

Let $\mathrm{lindisc}(K) = d$, then $dK + \mathbf{Z}^n$ covers the whole space. Hence $\mathrm{vol}(dK) \geq 1$ and so

(2) $$\mathrm{vol}(K) \geq (\mathrm{lindisc}(K))^{-1}.$$

The polar of the body K is also of interest in connection with discrepancy questions. Recall that the polar is defined by

$$K^* = \{x \in \mathbf{R}^n : x^T y \leq 1 \text{ for all } y \in K\}.$$

Now if $K = U_A$ then K^* is the convex hull of row vectors of A and their negatives, which we also denote by $\mathrm{conv}(\pm A)$.

We shall need a relation between the volume of a convex body and its polar. The following inequality for K is a convex body centrally symmetric with respect to the origin was proved by Blaschke (1917) and Santaló (1949). Let

$$c_n = \frac{\pi^{\frac{n}{2}}}{\Gamma(\frac{n}{2}+1)}$$

denote the volume of the unit ball in $\mathbf{R}^n$.

(3) $$\mathrm{vol}(K)\mathrm{vol}(K^*) \leq c_n^2.$$

The bound is attained by the ball. Note that $c_n^2 < (2\pi)^n/n!$. A lower bound for the product $\mathrm{vol}(K)\mathrm{vol}(K^*)$ of the form $(\mathrm{const})^n/n!$ was proved by Bourgain and Milman (1985). The old conjecture that this product is bounded from below by $4^n/n!$ (i.e., it is minimized by the cube) appears to be unsettled.

Combining (2) and (3), we obtain the following.

Lemma 1. $\mathrm{vol}(K^*) \leq c_n^2 \cdot \mathrm{lindisc}(K)^n$.

We shall apply this inequality in the next section to obtain an upper bound on the number of distinct rows of a matrix with given linear (or hereditary) discrepancy and given number of columns.

The inequality in Lemma 1 is certainly not sharp; equality in the Blaschke-Santaló inequality holds iff K is an ellipsoid, but then the translates of dK cannot cover the space without overlap. But it is interesting to point out that it is not too far from best possible. Considering any ball as K, we see that this cannot be replaced by any coefficient smaller than

$$\frac{2^n c_n}{n^{n/2}} \approx \frac{3.04^n}{n!}.$$

An even worse example is obtained if we consider the matrix A whose rows are all 0,1-vectors whose 1's are consecutive, and the body $K = U_A$. Then $K^* = \text{conv}(\pm A)$ consists of all vectors in $\mathbf{R}^n$ in which the sum of positive entries is at most 1 and the sum of negative entries is at least -1. Hence it is not difficult to determine that

$$\text{vol}(K^*) = \frac{\binom{2n}{n}}{n!} \approx \frac{4^n}{n!}.$$

As A is totally unimodular, its linear discrepancy is less than 1 and so the coefficient of $\text{lindisc}(K)^n$ in Lemma 1 could not be less than this number. It may be conjectured that this example is the worst (cf. the conjecture in the next section).

2. The Number of Rows of Matrices with Given Discrepancy

We determine the maximum number $h(n,d)$ of distinct edges in a hypergraph on n vertices with hereditary discrepancy d. Equivalently, $h(n,d)$ is the maximum number of distinct rows in a 0–1 matrix with hereditary discrepancy d. It seems that if we consider integral matrices instead of 0–1 matrices, then the problem becomes much harder. Let $f(n,d)$ denote the maximum number of distinct rows in an integral matrix with n columns and with hereditary discrepancy d. We give an upper bound for $f(n,d)$ and a construction which shows that, at least for large d, the bound is not too bad. We also formulate a conjecture on the exact value of $f(n,d)$.

Theorem 1. *For every integer $n \geq 1$ and half-integer $d \geq 1/2$,*

$$h(n,d) = 1 + n + \binom{n}{2} + \ldots + \binom{n}{4d}.$$

Proof I. By a theorem of Sauer (1972), if a hypergraph H on n vertices has more than $1 + n + \ldots + \binom{n}{k}$ distinct edges then there is a restriction of H to some $k+1$ vertices where all possible subsets of $k+1$ elements occur as edges. So if H has more than $1 + n + \binom{n}{2} + \ldots + \binom{n}{4d}$ edges then it has a restriction to some $4d+1$ vertices where all the possible subsets of this $(4d+1)$-element set occur as edges. Hence in any 2-coloration we find a monochromatic edge of size $2d+1$ which implies that $\text{herdisc}(H) > d$.

II. Now we give a construction of a hypergraph which has $1 + n + \binom{n}{2} + \ldots + \binom{n}{4d}$ edges on n vertices and has $\text{herdisc}(H) = d$. Take an ordered set of n elements. The edges of the hypergraph will be the subsets consisting not more than $2d$ intervals of the ordered set. Coloring the elements alternatingly red and blue we see that this hypergraph has discrepancy $\leq d$. Since every restriction is of the same structure, every restriction has discrepancy $\leq d$. An easy counting argument gives that the number of edges equals $1 + n + \binom{n}{2} + \ldots + \binom{n}{4d}$. □

Next we consider general integral matrices. To formulate our result, set

$$\phi(n,d) = \sum_{j=1}^{n} \binom{n}{j}\binom{n+j}{j}\binom{2d}{j}$$

and

$$\psi(n,d) = \sum_{j=1}^{n} \binom{n}{j} c_j^2 (2d)^j.$$

Then we have the following bounds:

Theorem 2. *For every integer $n \geq 1$ and half-integer $d \geq 1/2$,*

$$\phi(n,d) \leq f(n,d) \leq \psi(n,d).$$

Before proving the theorem, let us state slightly weaker upper and lower bounds, which are, however, easier to estimate. In particular, they show that for $d \geq n$, the value of $f(n,d)$ is between $(c_1 d)^n/n!$ and $(c_2 d)^n/n!$ with absolute constants c_1 and c_2.

Corollary.

$$\binom{n+2d}{n} \leq f(n,d) \leq \binom{2n+4\pi d}{n}.$$

Proof of Theorem 2. I. Let A be an integral matrix with n columns, m rows and with hereditary discrepancy d. Since every row of A is a lattice-point in $\text{conv}(\pm A), m \leq |\, Z^n \cap \text{conv}\,(\pm A)\,|$. We estimate the right hand side by the volume of $\text{conv}(\pm A)$. Let $Q = [0,1]^n$ be the unit cube, then the translate of Q by every integral vector in $\text{conv}(\pm A)$ is contained in $\text{conv}(\pm A) + Q$ and hence

$$m \leq |Z^n \cap \text{conv}(\pm A)| \leq \text{vol}(\text{con}(\pm A) + Q).$$

For any convex body B, $\text{vol}(B + Q)$ can be expressed as follows:

$$\text{vol}(B+Q) = \sum_T \text{vol}_T(B_T)$$

where T ranges over all subsets of the coordinates, B_T denotes the projection of B to the subspace spanned by these coordinates, and vol_T is the volume in this subspace.

Let us apply this formula with $B = \mathrm{conv}(\pm A)$. Notice that B_T is $\mathrm{conv}(\pm A_T)$, where A_T is the submatrix of A formed by the columns in T. Since together with A, A_T has hereditary discrepancy at most d and so linear discrepancy at most $2d$ by (1), we can use Lemma 1 and obtain that

$$m \leq \sum_T \mathrm{vol}_T(\mathrm{conv}(\pm A_T)) \leq \sum_{j=0}^{n} \binom{n}{j} c_j^2 (2d)^j.$$

II. To prove the lower bound, we give the following construction: let the rows of A be all integral vectors of length n in which the sum of positive entries is at most $2d$ and the sum of negative entries is at least $-2d$. The number of such vectors with exactly j positive entries is $\binom{n}{j} \times \binom{2d}{j} \times \binom{2d+n-j}{n-j}$. Summing this over all j, and using standard identities for binomial coefficients, we obtain that the number of rows of A is $\phi(n,d)$. □

We remark that there are other constructions of integral matrices with n columns, $\phi(n,d)$ rows and hereditary discrepancy d: let the rows of A' be all integral vectors $(a_1, \ldots, a_n)$ such that

$$\sum_{i=1}^{n+1} |a_i - a_{i-1}| \leq 4d$$

where $a_0 = a_{n+1} = 0$. This observation supports the following conjecture:

Conjecture. $f(n,d) = \phi(n,d)$.

References

[1] J.Beck (1981): Roth's estimate of the discrepancy of integer sequences is nearly sharp, *Combinatorica* **1** 319–325.

[2] Blaschke (1917):Über affine Geometrie VII: Neue Extremaleigenschaften von Ellipse und Ellipsoid, *Sitz. Ber. Akad. Wiss. Leipz. Math. Nat. Kl.* **69** 306–318.

[3] J. Bourgain and V.D. Milman (1985): Sections euclidiennes et volume des corps symétriques convexes dans $\mathbf{R}^n$, *C.R.Acad.Sci.Paris* **300** 435–437.

[4] A.Ghouila-Houri (1962): Caractérisation des matrices totalement unimodulaires, *C.R.Acad.Sci.Paris* **254** 1192-1194.

[5] I. Heller (1957): On linear systems with integral valued solutions, *Pac.J.Math.* **7** 1351–1364.

[6] L.Lovász, J.Spencer and K.Vesztergombi (1986): Discrepancy of set-systems and matrices, *Europ.J.Combin. (to appear)*

[7] N. Sauer (1972): On the density of families of sets, *J.Combin.Theory* A **13** 145–147.
[8] S. Shelah (1972): A combinatorial problem; stability and order for models and theories in infinitary languages, *Pac.J.Math.* **41** 247–261.

L. Lovász
L. Eötvös University
Department of Computer Science
Budapest, H-1088, Hungary

K.Vesztergombi
Budapest University of Technology
Faculty of Electrical Engineering
Department of Mathematics
Budapest, H-1111, Hungary

10. Spectral Studies of Automata

J.H. Loxton

1. Introduction

It is reasonable to argue that the simplest sequences are the periodic sequences. There are at least two ways to make this truism meaningful. The n-th term of a periodic sequence can be determined once we know the period and the integer n and this amounts to approximately $\log_2 n$ bits of information for large n. Also, the number of different words of length n in a periodic sequence is bounded. Periodic sequences minimise both of these characteristics. At the other end of the scale are the random sequences. We can take a random sequence to be one for which the most efficient way to find the n-th term is to make a list of the first n terms, requiring at least n bits of information. This implies another essential aspect of randomness that, for every n, the words of length n in the sequence are equidistributed.

Clearly, a deterministic process cannot produce a random sequence. Nevertheless, "random" numbers are computed in just this way and used with apparent success. (According to von Neumann, we therefore live in a state of sin.) We shall explore some aspects of this paradox, relating the complexity of the process with the randomness or otherwise of the sequences which it produces.

We concentrate here on the sequences generated by finite automata, that is computers without memory. These are the next simplest sequences after the periodic ones. There are several algebraic characterizations of "automatic sequences". What follows is directed instead at their analytic properties. We illustrate how the automaton can sometimes be identified by looking at the Fourier transform of the sequence and also why this idea does not yet give a completely satisfying answer to all the questions that can be asked.

A more detailed account of the matters raised will appear elsewhere.

2. Automata

Informally, a finite automaton is a finite state machine with an input and an output. The input symbols belong to a finite alphabet which we can take to be the set of digits to some base r. When a digit is read in, the machine moves from one state to another as prescribed by its transition function. When a string of digits is read in, the transition function operates sequentially on the digits, reading from right to left. The output depends on the final state after this processing is complete. In particular, let $a_s(n)$ be the output corresponding to the initial state s and the input string $(n)_r$ which is the string of digits of the integer n written in base r in the usual way. We shall say that any sequence $a(n)$ realized in this way as $a(n) = a_s(n)$ can be generated by a finite automaton in the base r. (It is more usual to require the automaton to read input strings from left to right, but this does not change the set of sequences which can be generated. See [5], page 18.)

The output sequences of finite automata give rise to functional equations. If $n = qr + t$ with $0 \leq t < r$, we can find $a_s(n)$ by first reading in the digit t, taking the machine from state s to state $t(s)$, say, and then reading in the digits of q. Thus

$$a_s(qr + t) = a_{t(s)}(q).$$

We can apply this recurrence to the generating function

$$f_s(z) = \sum_{n=0}^{\infty} a_s(n) z^n$$

giving

$$f_s(z) = \sum_{t=0}^{r-1} \sum_{q=0}^{\infty} a_s(qr+t) z^{qr+t} =$$

$$= \sum_{t=0}^{r-1} z^t \sum_{q=0}^{\infty} a_{t(s)}(q) z^{qr} = \sum_{t=0}^{r-1} z^t f_{t(s)}(z^r).$$

These equations have the shape

$$f(z) = A(z) f(z^r),$$

where $f(z)$ is the column vector whose components are the functions $f_s(z)$ and $A(z)$ is a matrix whose entries are rational functions of z. The generating function of a sequence generated by a finite automaton in the base r must be the first component, say, of a vector satisfying such a system of functional equations.

The first part of this argument shows that the block

$$a_s(qr) a_s(qr+1) \ldots a_s(qr + r - 1)$$

is completely determined by s and q, that is the output is generated by a substitution of length r. This gives an alternative description of the sequences generated by finite automata. (See [3], theorem 3.) In effect, a finite automaton can recognize certain patterns in the digits of the input. In general, the power series coefficients of solutions of the systems of functional equations developed above give a larger class of sequences, because they can recognize other functions of the input digits. However, if the coefficients can be taken in a finite field where all the arithmetic is performed, the resulting sequences are exactly those generated by finite automata in the base r. (See [1], section 7.)

Many examples are given in [3] and [1]; the following are simple but representative.

Example 1. The Thue sequence is 0110100110010110.... The n-th term, t_n say, is the parity of the number of ones in the binary representation of n, starting with $t_0 = 0$. Clearly,

$$t_{2n} = t_n, \; t_{2n+1} = 1 - t_n.$$

So t_n is generated by a finite automaton in base 2 with two states, 0 and 1, and two possible inputs, 0 and 1. The state of the machine changes whenever a 1 is read in as input. The initial state is 0 and the output is just the label of the final state. With this set-up, $a_0(n) = t_n$ and the generating function $f_0(z) = \sum t_n z^n$ satisfies

$$f_0(z) = (1 - z) f_0(z^2) + \frac{z}{1 - z^2}.$$

Nothing new arises from the initial state 1 because $a_1(n) = 1 - t_n$ and $f_1(z) = (1 - z)^{-1} - f_0(z)$. It seems that this sequence first found a serious use in work of Thue [16] on sequences without squares.

Example 2. Let $\sigma_r(n)$ denote the sum of the digits of n expressed in the base r in the usual way. For, $0 \le t < r$, $\sigma_r(qr + t) = \sigma_r(q) + t$, so the generating function $s_r(z) = \sum \sigma_r(n) z^n$ satisfies

$$s_r(z) = \sum_{t=0}^{r-1} \Big(s_r(z^r) z^t + \sum_{q=0}^{\infty} t z^{qr+t}\Big) =$$

$$= \frac{1 - z^r}{1 - z} s_r(z^r) + \frac{z - r z^r + (r-1) z^{r+1}}{(1 - z)^2 (1 - z^r)}.$$

This sequence is clearly not generated by a finite automaton because $\sigma_r(n)$ takes infinitely many values.

3. Change of Base

According to folklore, an irrational number with a simple definition should have a complicated decimal expansion. There are many natural conjectures along these lines, all of them apparently unattainable at present. For example, the decimal expansions of numbers such as $\sqrt{2}$, e and π should be normal. The normal numbers which have been written down explicitly are ad hoc constructions, mostly variations on the number 0.123456789101112..., and there is as yet no algebraic irrational which is known to be normal. In this context, the main theorem of [11] is a surprise. It goes some way towards proving that the decimal expansion of an algebraic irrational cannot be generated by a finite automaton. To this extent at least, the decimal expansions of algebraic irrationals are complicated.

Another natural conjecture is suggested in [1]. If $\{a_n\}$ is an infinite sequence of zeros and ones, then at least one of the numbers

$$\sum_{n=0}^{\infty} a_n 2^{-n}, \quad \sum_{n=0}^{\infty} a_n 3^{-n}$$

should be transcendental. A theorem of Cobham [2] settles this question in the context of finite automata. Suppose r and s are multiplicatively independent positive integers (that is, $\log r / \log s$ is irrational). Then a sequence which can be generated by a finite automaton in base r and in base s must be periodic. Finite automata are therefore base dependent. For some special digit patterns, there are stronger conclusions. For example, Senge and Straus have shown that there are only finitely many integers for which the sum of the digits in base r and in base s is less than a prescribed number. (Stewart [15] has given a quantitative version of this result and some extensions.) The extent to which these special results should generalize is not yet clear.

4. Spectral Analysis

We now wish to see how much Fourier analysis can tell us about the sequences of section 2. The simplest idea is to consider the Fourier transform of the sequence $\{a_n\}$, namely

$$\alpha(\theta) = \lim_{N\to\infty} \rho(N)^{-1} \sum_{n=0}^{N-1} a_n e(n\theta) \quad (e(\theta) = e^{2\pi i\theta}),$$

with an appropriate weighting factor ρ and a suitable specification of the limit.

In view of the functional equation for the generating function $f(z) = \sum a_n z^n$, it is more convenient to work with

$$\alpha(\theta) = \lim_{x \to 1} \rho((1-x)^{-1})^{-1} f(xe(\theta)).$$

This is substantially the same thing for the examples which occur here because of classical Abelian and Tauberian theorems. For example, if the a_n are bounded and one of the limits

$$\lim_{N \to \infty} \frac{1}{N} \sum_{n=0}^{N-1} a_n e(n\theta), \quad \lim_{x \to 1} (1-x) \sum_{n=0}^{\infty} a_n e(n\theta) x^n$$

exists, then so does the other, and they are equal because they are the Cesaro and Abel sums of the series $\sum a_n e(n\theta)$. In general, this equivalence holds if the a_n are bounded and ρ is regularly varying. (See [8], sections 5.12, 7.5 and 7.11.)

The following examples show what can be expected.

Example 1. Let $\tau_r(n)$ be the slightly modified Thue sequence defined by $\tau_r(n) = (-1)^s$ if n is a sum of s distinct powers of r and $\tau_r(n) = 0$ otherwise. The generating function is

$$t_r(z) = \sum_{n=0}^{\infty} \tau_r(n) z^n = \prod_{n=0}^{\infty} (1 - z^{r^n}).$$

(The function $f_0(z)$ in example 1 of section 2 is just $\frac{1}{2}((1-z)^{-1} - t_2(z))$ and its spectrum has an extra atom at $\theta = 0$ corresponding to the pole of $\frac{1}{2}(1-z)^{-1}$ at $z = 1$.) Clearly, everything converges for $|z| < 1$ and $t_r(z)$ has the unit circle as a natural boundary. We set $z = xe(\theta)$ and consider the behaviour of $t_r(z)$ as $x \to 1$. Observe that

$$\int_0^1 |t_r(xe(\theta)|^2 d\theta = \sum_{\tau_r(n) \neq 0} x^{2n} = \prod_{n=0}^{\infty} (1 + x^{2r^n}) = (1-x)^{\frac{-\log 2}{\log r}} e^{O(1)}$$

as $x \to 1$. (The estimate can be obtained by following the method of [4].) From all this,

$$\lambda_r(\theta) = \lim_{x \to 1} (1-x)^{\frac{\log 2}{2 \log r}} t_r(xe(\theta))$$

is in L^2 and the Riesz product

$$|\lambda_r(\theta)|^2 = \prod_{n=0}^{\infty} (1 - \cos r^n \theta)$$

is a continuous singular measure on R/Z. (See [17], pages 149-150 and 208-209.) This is a particular case of the results of Keane [10]. Unfortunately, it is difficult to picture the behaviour of the Riesz product. The estimates can be carried a step further, as follows. From [4],

$$\log t_r(e^{-t}) = \frac{(\log t^{-1})^2}{2\log r} - \frac{1}{2}\log t^{-1} + O(1) \text{ as } t \to 0.$$

It follows that if $\theta = q/r^k$ with q not divisible by r, then

$$\log t_r(e^{-t}e(q/r^k)) = \log t_r(e^{-tr^k}) + O(1) = \\ = -\frac{(\log t^{-1})^2}{2\log r} + (k - \frac{1}{2})\log t^{-1} + O(1) \text{ as } t \to 0.$$

Now suppose θ is not an r-adic rational. We can proceed as in [12] to get the approximation

$$\log t_r(e^{-t}e(\theta)) = \sum_{n=0}^{N-1} \log(1 - e^{-tr^n} - 2\pi i\|r^n\theta\|) + O(\log t^{-1}),$$

where $N = [\log t^{-1}/\log r]$ and $\|r^n\theta\|$ denotes the signed distance from $r^n\theta$ to the nearest integer. For almost all θ, this is $O(\log t^{-1})$. The exceptions occur when the representation of θ in base r has arbitrarily long blocks of zeros, but even in such a case we can show

$$\log|t_r(e^{-t}e(\theta))| = -\frac{(\log t^{-1})^2}{2\log r} + \psi(t)\log t^{-1}$$

where $\psi(t) \to \infty$ as $t \to 0$. These observations reveal the r-adic nature of the spectrum. Figures 1 and 2 illustrate this for $r = 2$. Figure 1 is an approximation to the graph of the Fourier transform of τ_2, namely $\lambda(\theta) = \sum_{n<20000} \tau_2(n)e(n\theta)$, and figure 2 is the graph of $-\log\lambda(\theta)$ clearly showing the peaks at the diadic rationals $p/2^4$.

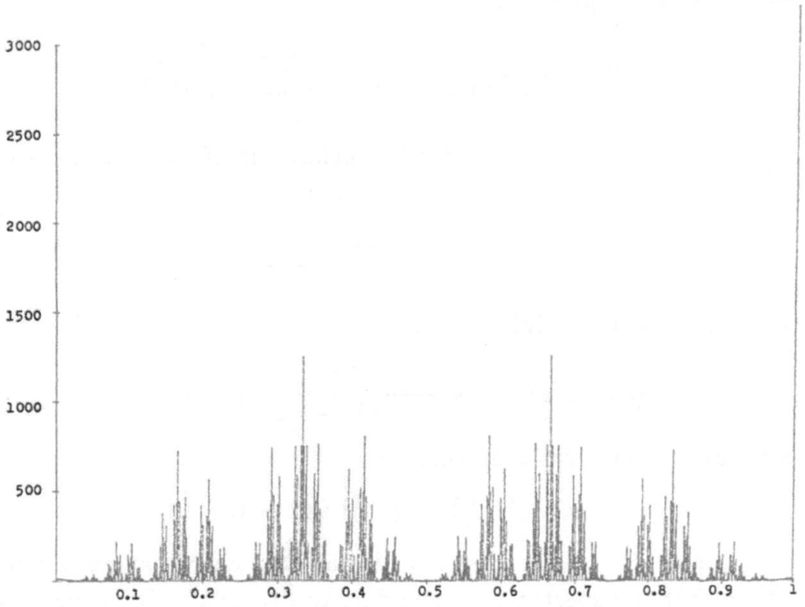

Figure 1

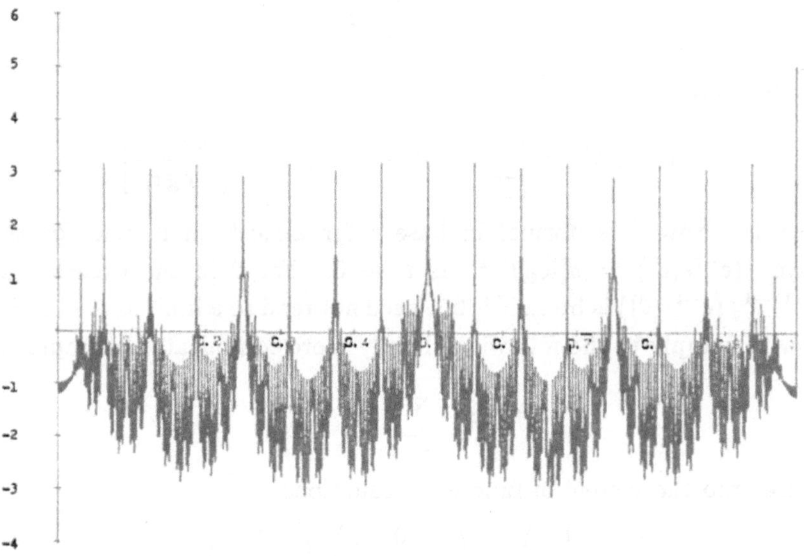

Figure 2

Example 2. Let $\sigma_r(n)$ denote the sum of the digits of n in base r and set $s_r(z) = \sum \sigma_r(n) z^n$ as in section 2. By iterating the functional equation for $s_r(z)$, we obtain the representation

$$s_r(z) = \frac{1}{1-z}\{\frac{z}{1-z} - (r-1)\sum_{j=1}^{\infty} \frac{z^{r^j}}{1-z^{r^j}}\}.$$

Again set $z = e^{-t}e(\theta)$ and consider the behaviour of $s_r(z)$ as $t \to 0$. By arguments like those used in example 1, we obtain

$$s_r(e^{-t}) \sim \frac{1}{2\log r}\frac{\log t^{-1}}{t} \text{ as } t \to 0;$$

if $\theta = q/r^k$ with q not divisible by r, then

$$s_r(e^{-t}e(q/r^k)) \sim \frac{r^{1-k}}{(e(q/r^k)-1)t} \text{ as } t \to 0,$$

and if θ is not an r-adic rational, then

$$s_r(e^{-t}e(\theta)) = o(t^{-1}) \text{ as } t \to 0.$$

Thus $s_r(z)$ yields discrete spectrum with non-zero charge at each r-adic rational. Note that the spectra corresponding to two relatively prime bases are disjoint, except for the charge at $\theta = 0$. This is similar to the result of Kamae [9] dealing with the spectral measure of the function $e(\sigma_r(n))$ instead of our $\sigma_r(n)$.

Example 3. In [12], Mahler treats the Fredholm series

$$f(z) = \sum_{n=0}^{\infty} z^{r^n}$$

and shows that

$$f(e^{-t}e(\theta)) = \sum_{n=0}^{N-1} e(r^n\theta) + O(1),\ N = \left[\frac{\log t^{-1}}{\log r}\right]$$

as $t \to 0$. Now θ is normal in base r for almost all θ, and for these θ we have $f(e^{-t}e(\theta)) = o(\log t^{-1})$ as $t \to 0$. For θ in the exceptional set, $(\log t^{-1})^{-1} f(e^{-t}e(\theta))$ is bounded, but need not tend to a limit as $t \to 0$. Similar observations apply to the following slightly more complicated example. Let

$$g(z) = z + \sum_{n=0}^{\infty}\sum_{m=0}^{n} z^{r^n + r^m}$$

which fits into the system of functional equations

$$\begin{pmatrix} 1 \\ f(z) \\ g(z) \end{pmatrix} = \begin{pmatrix} 1 & 0 & 0 \\ z & 1 & 0 \\ 0 & z & 1 \end{pmatrix} \begin{pmatrix} 1 \\ f(z^r) \\ g(z^r) \end{pmatrix}.$$

With $z = e^{-t}e(\theta)$ and N as before, we see that the terms of the series with $n \geq N$ contribute $O(1)$ and we can replace z by $e(\theta)$ in the remaining terms with total error $O(N)$. So

$$g(e^{-t}e(\theta)) = \sum_{n=0}^{N-1} \sum_{m=0}^{n} e((r^n + r^m)\theta) + O(N), \; N = \left[\frac{\log t^{-1}}{\log r}\right]$$

as $t \to 0$. In this case $g(e^{-t}e(\theta)) = o((\log t^{-1})^2)$ if θ is normal in base r.

5. Applications

(1) Distributions. Weyl showed that if u_n is any increasing sequence of integers, then $u_n\theta$ is uniformly distributed mod 1 for almost all θ. For some special sequences, the set of normality can be determined more precisely. Recently, Mauduit [13] has given a theorem which includes and generalizes many earlier results. A similar result can be obtained using the ideas of the previous section and we give it next. In the course of the argument, we will obtain some general information on the growth of our generating functions near their natural boundaries.

Suppose the sequence a_n is generated by a finite automaton and that there are infinitely many solutions to the equation $a_n = a$, say $n = u_1, u_2, \ldots$. If $\log u_n$ grows more slowly than any power of n, then $u_n\theta$ is uniformly distributed mod 1 for every irrational θ. Some restriction on the growth of the sequence is necessary because, for example, $2^n\theta$ is uniformly distributed mod 1 if and only if θ is normal in base 2.

We may suppose that a_n is a sequence of non-negative integers generated by a finite automaton in base r. Consider the associated system of functional equations

$$f(z) = A(z)f(z^r)$$

derived as in section 2. Thus each row of the matrix $A(z)$ contains some of the entries $1, z, \ldots, z^{r-1}$ and zeros, one of the components of $f(z)$ is $\sum a_n z^n$, and the other components are sections of this series of the shape $\sum a_{r^k n + t} z^n$. We take the smallest possible system with these properties. We can relabel the components of $f(z)$ so that the non-negative matrix $A(1)$ is in normal form with the block decomposition

$$A(1) = \begin{pmatrix} A_1 & & & \\ A_{21} & A_2 & & \\ \vdots & \vdots & \ddots & \\ A_{s1} & A_{s2} & \ldots & A_s \end{pmatrix}, \quad f(z) = \begin{pmatrix} f_1(z) \\ f_2(z) \\ \vdots \\ f_s(z) \end{pmatrix},$$

where the diagonal blocks $A_1, \ldots, A_s$ are irreducible and their dominant eigenvalues increase down the diagonal, the blocks $A_{21}, \ldots, A_{s,s-1}$ are non-zero and the original generating function is a component of $f_s(z)$. (See [6], page 90.) Let ρ denote the dominant eigenvalue of $A(1)$. There are now two possibilities. This dichotomy makes essential use of the fact that the entries of $A(1)$ are zeros and ones.

Type 1: $\rho = 1$. If all the diagonal blocks of $A(1)$ are scalar matrices, then $\rho = 1$ and

$$\sum_{n<N} a_n \sim c\left(\frac{\log N}{\log r}\right)^q \quad (N \to \infty),$$

$$f_s(e^{-t}) \sim c\left(\frac{\log t^{-1}}{\log r}\right)^q \quad (t \to 0),$$

where q is an integer in the range $1 \le q \le s-1$ and c is a positive constant. This case is typified by example 3 of section 4.

Type 2: $\rho > 1$. If one of the diagonal blocks of $A(1)$ is not a scalar matrix, then $\rho > 1$ and

$$\sum_{n<N} a_n \sim cN^{\log \rho/\log r},$$

$$\sum_{n<N} a_n e(n\theta) = o(\sum_{n<N} a_n) \quad (N \to \infty)$$

for every irrational θ. Here, the analysis is similar to that in example 2 of section 4.

The assertions about the uniform distribution of $u_n\theta$ follow by applying Weyl's criterion with the above estimates. For type 2, $u_n\theta$ is uniformly distributed mod 1 for every irrational θ, while for type 1, $\log u_n \gg n^{1/q}$ and the set of bad distribution can be much larger. In [13], the growth condition on u_n is interpreted in terms of the graph associated with the automaton.

(2) Discrimination. The calculations suggest an alternative analytic proof of Cobham's theorem mentioned in section 3. It should be possible to discern the base used in a finite automaton in the behaviour of the Fourier transform of the sequence generated by it. In many cases, as in [9], the examples of section 4 and automata of type 1, this can indeed be verified. The general case is not quite so transparent, but has been carried through by Queffelec [14] using the language of ergodic theory. It appears that the limit distributions defined by the generating functions on their natural boundaries are generalized Riesz products and so two systems with multiplicatively independent bases are spectrally independent.

(3) Discovery. Can we decide whether or not a sequence is generated by a finite automaton by looking at its Fourier transform? An interesting test comes from a question asked by Odlyzko and Stanley. (See [7], problem E1O.) The famous $r_3(n)$ is the least r such that any sequence $1 \le u_1 < u_2 < \ldots < u_r \le n$ of r numbers not exceeding n must contain three terms forming an arithmetic progression. The best bounds so far obtained are

$$ne^{-c\sqrt{\log n}} < r_3(n) < \frac{c'n}{\log\log n}.$$

The obvious way to construct sequences not containing three terms in arithmetic progression is to use the greedy algorithm, but the resulting sequences are not very dense. For example, this process yields the sequence

$$A(1): \ 0,1,3,4,9,10,12,13,27,28,30,31,36,37,39,40,81,82,84,85,\ldots.$$

The elements of $A(1)$ are just the integers whose base 3 representations do not contain the digit 2, and the number of terms in $A(1)$ not exceeding n is approximately $n^{\log 2/\log 3}$. Let $A(m)$ be the sequence obtained by starting with $u_0 = 0$ and $u_1 = m$ and then taking each subsequent u_{n+1} as the least number greater than u_n so that $\{u_0, u_1, \ldots, u_{n+1}\}$ does not contain three terms in arithmetic progression. If m is a power of three, or twice a power of three, the numbers of the sequence $A(m)$ can be described simply in terms of their base 3 representations, so $A(m)$ can be produced by a finite automaton is base 3. For example, consider

$$A(6): \ 0,6,7,9,10,15,16,19,27,33,34,36,37,42,43,46,81,87,88,90,\ldots.$$

The graph in figure 3 is an approximation to the Fourier transform of $A(6)$. ($\alpha_6(\theta) = \sum_{n<1000} e(u_n\theta)$ is plotted against θ, u_n being the n-th term of $A(6)$.) The discrete spectrum of $A(6)$ located at the 3-adic rationals is clearly visible, in accord with the fact that the sequence can be produced by a finite automaton in base 3.

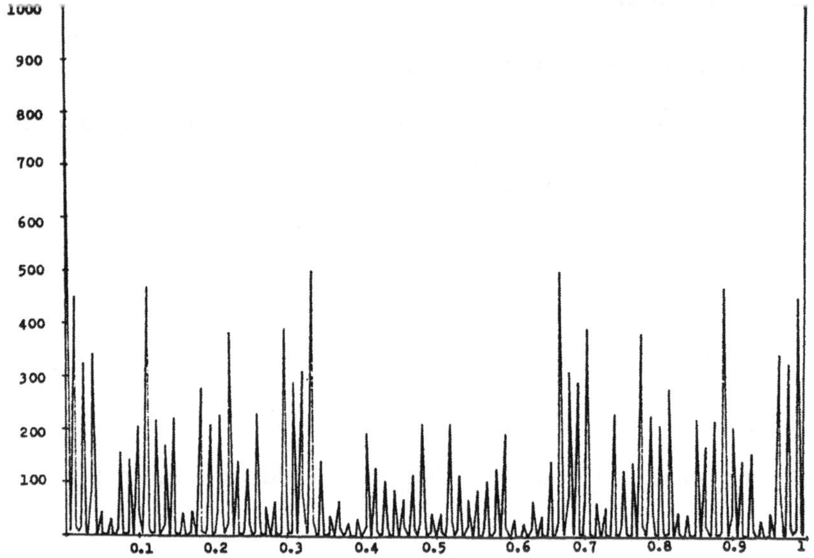

Figure 3

The problem raised by Odlyzko and Stanly is to explain the behaviour of $A(m)$ for other m. These sequences seem to have the same rate of growth as $A(1)$, but they are more erratic. For example,

$$A(4):\ 0,4,5,7,11,12,16,23,26,31,33,37,38,44,49,56,73,78,80,85,\ldots$$

has the Fourier transform shown in figure 4. (Again, this is based on the first 1000 terms.) This appears quite different to figure 3. However, example 1 of section 4 shows that the spectrum of a sequence generated by a finite automaton need not have infinitely many atoms, so we cannot make any easy conclusions. More delicate analysis beyond that of the preceding paragraphs, or more delicate experiments, seem to be indicated.

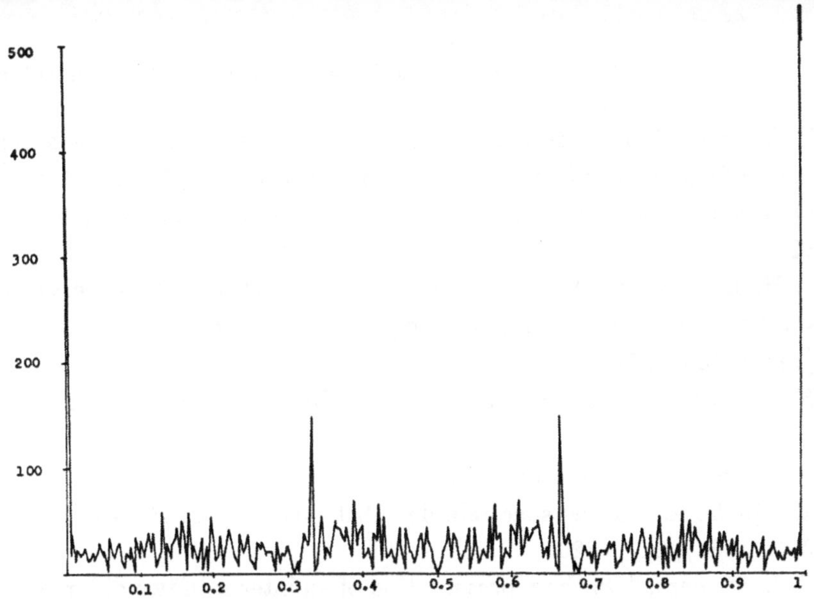

Figure 4

References

[1] G. Christol, T. Kamae, M. Mendes-France and G. Rauzy, "Suites algébriques, automates et substitutions." *Bull. Soc. Math. France* **108** (1980), 401 - 419.

[2] A. Cobham, "On the base dependence of sets of numbers recognizable by finite automata." *Mathematical Systems Theory* **3** (1969), 186 - 192.

[3] A. Cobham, "Uniform tag sequences." *Mathematical Systems Theory* **6** (1972), 164 - 192.

[4] N. G. de Bruijn, "On Mahler's partition problem." *Proc. Nederl. Akad. Wetensch.* **51** (1948), 659 - 669. (=*Indag. Math.* **10** (1948), 210 -220.)

[5] S. Eilenberg, *Automata, languages and machines.* Vol. A. (Academic Press, 1974.)

[6] F. R. Gantmacher, *Applications of the theory of matrices.* (Wiley, 1959.)

[7] R. K. Guy, *Unsolved problems in number theory.* (Springer, 1981.)

[8] G. H. Hardy, *Divergent series* (Oxford, 1949.)

[9] T. Kamae, "Mutual singularity of spectra of dynamical systems given by *sums of digits* to different bases." *Astérisque* **49** (1977), 109 - 114.

[10] M. Keane, "Strongly mixing g-measures." *Inv. Math.* **16** (1972), 309 - 324.

[11] J. H. Loxton and A. J. van der Poorten, "Arithmetic properties of the solutions of a class of functional equations." *J. reine angew. Math.* **330** (1982), 159 - 172.

[12] K. Mahler, "On a special function." *J. Number Theory* **12** (1980), 20 - 26.

[13] C. Mauduit, "Automates finis et équirépartition modulo 1." *C. R. Acad. Sc. Paris* **299** (1984), 121 - 123.

[14] M. Queffelec, "Etude spectrale de substitutions." *C. R. Acad. Sc. Paris* **297** (1983), 317 - 320.

[15] C. L. Stewart, "On the representation of an integer in two different bases." *J. reine angew. Math.* **319** (1980), 63 - 72.

[16] A. Thue, "Über die gegenseitige Lage gleicher Teile gewisser Zeichenreichen." *Christiania Vidensk. Selsk. Skr.* (1912), no. 1. (= *Selected mathematical papers*, Universitetsfolaget, Oslo, 1977, 413 - 477.)

[17] A. Zygmund, *Trigonometric series.* **1.** (Cambridge, 2nd edition, 1979.)

J.H. Loxton
School of Mathematics and Physics
Macquarie University
New South Wales, 2109
AUSTRALIA

11. A Diophantine Problem

M. Mendes France

1. The Norm of a Sequence

Let $\mathbf{v} = (v_n)$, $n \geq 0$, be an infinite sequence (mod 1). Let $p \in [1, +\infty]$. The symbol $|..|$ denotes the "norm" on the torus $\mathbf{R}/\mathbf{Z}$, i.e the distance to the nearest integer. We define the "norm" of the sequence $\mathbf{v}$

$$\|\mathbf{v}\|_p = \limsup_{N\to\infty} \left(\frac{1}{N} \sum_{n=0}^{N-1} |v_n|^p \right)^{1/p} \quad \text{for } 1 \leq p < \infty,$$

and

$$\|\mathbf{v}\| = \|\mathbf{v}\|_\infty = \limsup_{N\to\infty} |v_N|.$$

Let $q \geq 2$ be an integer. We consider the sequence $q = (q^n)$ with which we associate the convolution operator

$$\mathbf{v} \to \mathbf{v} * \mathbf{q}$$

defined by

$$(\mathbf{v} * \mathbf{q})_n = v_n + q v_{n-1} + q^2 v_{n-2} + \ldots + q^n v_0.$$

We are interested in the "eigenvalue problem" associated with the operator $\mathbf{q}$ and prove that

$$\lim_{q\to\infty} \inf_{\xi\in\mathbf{R}} q \|\mathbf{v} * \mathbf{q} - \xi \mathbf{q}\|_p = \|\mathbf{v}\|_p. \tag{1}$$

We shall then apply this result to the special sequence $v_n = x\theta^n$ where $x \neq 0$ and $\theta > 1$.

2. A Diophantine Inequality and Some Consequences

Let T to be shift operator $T(u_n) = (u_{n+1})$. The following result which we will prove in paragraph 3 is an extension of an inequality established in [1].

Theorem 1. *Given the real sequence* $\mathbf{u}$ *and the integer* $q \geq 2$, *there exists* $\omega \in \mathbf{R}$ *such that*

$$(q-1)\|\mathbf{u}-\omega\mathbf{q}\|_p \leq \|T\mathbf{u}-q\mathbf{u}\|_p \leq (q+1)\|\mathbf{u}-\omega\mathbf{q}\|_p.$$

For all real ξ

$$\|T\mathbf{u}-q\mathbf{u}\|_p \leq (q+1)\|\mathbf{u}-\xi\mathbf{q}\|_p.$$

Consider the sequence $\mathbf{v}$ defined by $v_0 = u_0$ and $T\mathbf{v} = T\mathbf{u} - q\mathbf{u}$. The operator $T-q$ is invertible and it can easily be checked that

$$\mathbf{u} = (T-q)^{-1}\mathbf{v} = \mathbf{v} * \mathbf{q}.$$

Theorem 1 asserts the existence of $\omega \in \mathbf{R}$ such that

$$(q-1)\|\mathbf{v}*\mathbf{q}-\omega\mathbf{q}\|_p \leq \|\mathbf{v}\|_p \leq (q+1)\|\mathbf{v}*\mathbf{q}-\omega\mathbf{q}\|_p$$

which implies equality (1).

As a consequence of the above result we shall obtain the following two theorems. For all real x let $[[x]]$ be the nearest integer and

$$((x)) = x - [[x]]$$

with the convention $((\frac{1}{2})) = \frac{1}{2}$.

Theorem 2. *Suppose* $\|\mathbf{v}\| > 0$. *If for all positive integers* $k < (2(q-1)/\|\mathbf{v}\|)^2$ *there exists an interval* $I(k)$ *of length strictly larger than* $16\log q/\log\frac{2(q-1)}{\|\mathbf{v}\|}$ *which contains no terms of the sequence* $k\mathbf{v}*\mathbf{q}$ *then the real number*

$$\sum_{n=1}^{\infty}\frac{((v_n))}{q^n}$$

is rational.

If one applies this result to the sequence $v_n = x\theta^n$ where $x \neq 0$ and $\theta > 1$, we see that under the above conditions,

$$\sum_{n=1}^{\infty}((x\theta^n))q^{-n}$$

is rational, a result to be compared to one of Salem's observations according to which the entire series

$$\sum_{n=1}^{\infty}((x\theta^n))z^n$$

is a rational function of z provided θ is either a PV number or a Salem number [5].

Theorem 2 is meaningless if $\|\mathbf{v}\| = 0$. Let $\eta > 0$. A real sequence $\mathbf{v}$ is said to be η-dense (mod 1) if every interval the length of which is strictly larger than η contains infinitely many terms of $\mathbf{v}$.

Theorem 3. *Let* $\mathbf{v}$ *be a sequence such that* $\|\mathbf{v}\| = 0$. *Then either one of the two alternatives hold.*

1. *There exists an integer* $l \neq 0$ *such that* $\|l\mathbf{q} * \mathbf{v}\| = 0$;
2. *For all* $\eta > 0$, *there exists an integer* $k < q^{6q/\eta}$ *such that the sequence* $k\mathbf{v} * \mathbf{q}$ *is* η *dense (mod 1).*

Corollary. *Let* $\theta > 2$ *be a noninteger real number and suppose* $x \neq 0$. *If*

$$\limsup_{n\to\infty} x\theta^n = 0,$$

then either there exists a nonzero integer l *such that*

$$\limsup_{n\to\infty} \left| \frac{lx}{\{\theta\}} \theta^n \right| = 0,$$

or for all $\eta > 0$ *there exists an integer* $k < \theta^{6\theta/\eta}$ *for which the sequence* $k\frac{x}{\{\theta\}}\theta^n$ *is* η*-dense (mod 1).*

This result is clearly a consequence of Theorem 3 with $v_n = x\theta^n$ and $q = [\theta]$.

Pisot's conjecture [4] states that if $|x\theta^n|$ tends to o then θ is a PV number and $x \in \mathbf{Q}(\theta)$. If the conjecture holds true, then $\frac{x}{[\theta]}\theta^n$ has finitely many limit points, all of which are rational hence there exists a nonzero integer l such that $l\frac{x}{\{\theta\}}\theta^n$ tends to zero (mod 1). The second alternative would thus be impossible.

3. Proof of Theorem 1

We first observe that for all real ξ

$$u_{n+1} - qu_n = (u_{n+1} - \xi q^{n+1}) - q(u_n - \xi q^n)$$

hence by Hölder's inequality $(1 \le p < \infty)$:

$$|u_{n+1} - qu_n| \le (l+q)^{1/p'}(|u_{n+1} - \xi q^{n+1}|^p + q|u_n - \xi q^n|^p)^{1/p},$$

where $\frac{1}{p} + \frac{1}{p'} = 1$. Hence

$$\sum_{n=0}^{N-1} |u_{n+1} - qu_n|^p \le (1+q)^{p/p'} \left[\sum_{n=0}^{N-1} |u_n - \xi q^n|^p (1+q) + O(1) \right],$$

thus

$$\|T\mathbf{u} - q\mathbf{u}\|_p \leq (1+q)\|\mathbf{u} - \xi\mathbf{q}\|_p.$$

The inequality obviously extends to the case where $p = \infty$. We now define

$$w = \sum_{n=1}^{\infty} \frac{((u_n - qu_{n-1}))}{q^n},$$

where we assume $u_0 = 0$ with no less of generality. Then

$$\omega q^n = ((u_1 - qu_0))q^{n-1} + \ldots + ((u_n - qu_{n-1})) + \sum_{l=n+1}^{\infty} \frac{((u_l - qu_{l-1}))}{q^{l-n}}$$

$$\equiv u_n + \sum_{l=n+1}^{\infty} \frac{((u_l - qu_{l-1}))}{q^{l-n}} \pmod 1.$$

$$|\omega q^n - u_n| \leq \sum_{l=n+1}^{\infty} \frac{|u_l - qu_{l-1}|}{q^{l-n}}$$

(2)
$$\leq \sum_{l=n+1}^{\infty} \frac{1}{q^{l-n}} \sup_{l\geq n+1} |u_l - qu_{l-1}|$$

$$\leq \frac{1}{q-1} \sup_{1\leq n+1} |u_l - qu_{l-1}|,$$

thus

$$\|\omega\mathbf{q} - \mathbf{u}\| \leq \frac{1}{q-1}\|T\mathbf{u} - q\mathbf{u}\|.$$

We now prove the inequality for the p-norm, $p < \infty$. Inequality (2) can be written as

$$|\omega q^n - u_n| \leq \sum_{l\geq n+1} \frac{|u_l - qu_{l-1}|}{(q^{l-n})^{1/p}} \frac{1}{(q^{l-n})^{1/p'}},$$

hence

$$|\omega q^n - u_n| \leq \left[\sum_{l\geq n+1} \frac{|u_l - qu_{l-1}|^p}{q^{l-n}}\right]^{1/p} \left(\sum \frac{1}{q^{l-n}}\right)^{1/p'}$$

$$\leq \frac{1}{(q-1)^{1/p'}} \left[\sum_{l\geq n+1} \frac{|u_l - qu_{l-1}|^p}{q^{l-n}}\right]^{1/p}.$$

Then

$$\frac{1}{N}\sum_{n=1}^{N} |\omega q^n - u_n|^p \leq \frac{1}{(q-1)^{p/p'}} \sum_{l=1}^{\infty} \frac{1}{q^l} \cdot \frac{1}{N} \sum_{n=1}^{N} |u_{l+n} - qu_{l+n-1}|^p$$

$$\|\omega\mathbf{q} - \mathbf{u}\|_p^p \leq \frac{1}{(q-1)^{p/p'}} \cdot \frac{1}{(q-1)} \|T\mathbf{u} - q\mathbf{u}\|_p^p,$$

finally

$$\|\omega \mathbf{q} - \mathbf{u}\|_p \leq \frac{1}{q-1}\|T\mathbf{u} - q\mathbf{u}\|_p.$$

□

4. Proof of Theorem 2

The proof is based on a beautiful result of Mahler [2] which we state here as a lemma.

Lemma. *Let α be an irrational number. For all integer $\nu \geq 1$ there exists an integer $k < q^{2q^\nu + 2\nu - 1}$ such that the expansion of $k\alpha$ in base q contains each bloc of ν digits infinitely many times.*

In other terms, each interval $(i/q^\nu, (i+1)/q^\nu)$, $i = 0, 1, \ldots, q^\nu - 1$ contains infinitely many terms of the sequence $k\alpha q^n$. The sequence $k\alpha\mathbf{q}$ is thus $(2q^{-\nu})$-dense (mod 1).

Let $\eta \in]0,1[$. Choose the integer ν so that

$$\frac{2}{q^\nu} \leq \eta < \frac{2}{q^{\nu-1}}.$$

Then

$$2q^\nu + 2\nu - 1 < 6q/\nu.$$

Mahler's result hence states there exists an integer $k < q^{6q/\eta}$ such that $k\alpha\mathbf{q}$ is η-dense (mod 1). Let $\mathbf{v} = (v_n)$ be a sequence such that

$$\|\mathbf{v}\| = \limsup_{n \to \infty} |v_n| > 0.$$

Theorem 1 shows that the number

$$\omega = \sum_{n=1}^{\infty} \frac{((v_n))}{q^n}$$

satisfies the condition

$$\|\mathbf{v} * \mathbf{q} - \omega\mathbf{q}\| \leq \frac{\|\mathbf{v}\|}{q-1}. \tag{3}$$

Let us suppose that ω is irrational. There exists an integer $k = k(\nu) < q^{2q^\nu + 2\nu - 1}$ such that $k\omega\mathbf{q}$ is $(2q^{-\nu})$-dense (mod 1). Inequality (3) shows that $k\mathbf{v} * \mathbf{q}$ is $(2q^{-\nu} + k(q-1)^{-1}\|\mathbf{v}\|)$-dense (mod 1). We now choose ν in order to optimize

$$\frac{2}{q^\nu} + \frac{k}{q-1}\|\mathbf{v}\| < \frac{2}{q^\nu} + \frac{1}{q-1}\|\mathbf{v}\| q^{2q^\nu + 2\nu - 1},$$

namely

$$\frac{2}{q^\nu} = \frac{1}{q-1}\|\mathbf{v}\| q^{2q^\nu+2\nu-1},$$

hence

$$2q^\nu + 3\nu - 1 = \frac{\log \frac{2(q-1)}{\|\mathbf{v}\|}}{\log q}.$$

Now

$$2q^\nu \le 2q^\nu + 3\nu - 1 \le 4q^\nu,$$

hence

$$q^\nu = c\frac{|\log \frac{2(q-1)}{\|\mathbf{v}\|}|}{\log q},$$

where

$$\frac{1}{4} \le c \le \frac{1}{2}.$$

Finally

$$\frac{2}{q^\nu} = \frac{2\log q}{c\log \frac{2(q-1)}{\|\mathbf{v}\|}} \le \frac{8\log q}{\log \frac{2(q-1)}{\|\mathbf{v}\|}}.$$

Thus, if ω is irrational, there exists an integer

$$k < q^{4q^\nu} < \left(\frac{2(q-1)}{\|\mathbf{v}\|}\right)^2$$

such that the sequence $k\mathbf{v} * \mathbf{q}$ is η-dense (mod 1) where

$$\eta = \frac{2}{q^\nu} + \frac{k\|\mathbf{v}\|}{q-1} < 16\frac{\log q}{\log \frac{2(q-1)}{\|\mathbf{v}\|}}.$$

This contradicts the assumption of Theorem 2 hence ω is rational.

5. Proof of Theorem 3

We now assume $\|\mathbf{v}\| = 0$. Then

$$\|\mathbf{q} * \mathbf{v} - \omega\mathbf{q}\| = 0.$$

If ω is rational, say $\omega = a/l$, then $\omega\mathbf{q}$ is ultimately periodic (mod 1) and takes its values in the finite set $\{0, \frac{1}{l}, \frac{2}{l}, \ldots, \frac{l-1}{l}\}$. The sequence $l\omega\mathbf{q}$ is the null sequence, thus $l\mathbf{q} * \mathbf{v}$ tends to 0. If ω is irrational, Mahler's result shows that there exists an integer $k < q^{6q/\eta}$ such that $k\omega\mathbf{q}$ is η-dense (mod 1). The sequence $k\mathbf{q} * \mathbf{v}$ is then also η-dense (mod 1).

References

[1] J. Coquet, M. Mendes France. Suites à spectre vide et suites pseudo aléatoires, *Acta Arith.* **32**, 1977, 99 - 106.

[2] K. Malher, Arithmetical properties of the digits of multiples of an irrational number, *Bull. Aust. Math. Soc.* bf 8, 1973, 191 - 203.

[3] M. Mendes France, A diophantine inequality, *Sitzungsberichte der Österreichischen Akademie der Wissenschaften, in honour of E. Hlawka, Abt. II, Math. Phys.* **195**, 1986, 105 - 108.

[4] Ch. Pisot, La répartition 1 et les nombres algébriques, *Ann. Scuola Norm. Sup. Pisa*, **7**, 1938, 205 - 248.

[5] R. Salem, Power series with integral coefficients, *Duke Math. J.* 1945, **12**, 153 - 171.

M. Mendes France
Départment de Mathématiques
Université Bordeaux
F-33405 Talence

12. A Note on Boolean Dimension of Posets

J. Nešetřil and P. Pudlák

Abstract

We present a universal upper bound for the boolean dimension of posets. We prove that this bound is asymptotically best possible.

Introduction and Statement of Results

It is well known that every poset $P = (X, <)$ can be expressed as an intersection of linear orders. Putting otherwise for every P there are labellings $l_1, \ldots, l_k : X \to N$ such that $x < y$ if $l_i(x) < l_i(y)$ for $i = 1, \ldots, k$. The minimal k with this property is called the (Dushnik – Miller) dimension dim P of P. [4] is a very nice survey on the dimension of posets. Particularly, recall that dim $P_n \leq \frac{n}{2}$ (P_n denotes a poset with n points), while equality is achieved only for the "standard example": $X = \{1, 2, \ldots, n, 1', \ldots, n'\}$

$$i < j' \Leftrightarrow i \neq j.$$

Motivated by applications in data – base theory the following more powerful type of representation was introduced in [2]:

A *boolean representation* of a poset $P = (X, <)$ is a set $l_1, \ldots, l_k$ of labellings $X \to \mathbf{N}$ and a boolean formula $\varphi(a_1, \ldots, a_k)$ such that the following is true: $x < y$ iff $\varphi(a_1, \ldots, a_k) = 1$ where

$$a_i = \begin{cases} 1 & \text{if } l_i(x) < l_i(y), \\ 0 & \text{if } l_i(x) > l_i(y). \end{cases}$$

The minimal k with the above property is called the boolean dimension $\dim_B(P)$ of P

Clearly $\dim_B(P) \leq \dim(P)$ as the Dushnik — Miller dimension corresponds to the formula $\varphi = x_1 \wedge \ldots \wedge x_k$. One may prove that the boolean dimension of the standard example is ≤ 4. It seems difficult to construct posets with high boolean dimension. The purpose of this note is to prove the following two statements:

Proposition 1.

$$\dim_B(P_n) \le c \log n$$

for a suitable constant c (independent on n).

Proposition 2. *There exists a positive constant c' such that for every n there are posets P_n with $\dim_B(P_n) \ge c' \log n$.*

Note that the proof of Proposition 1 is constructive and yields a polynomial algorithm for boolean representation. Our proof of Proposition 2 is nonconstructive and presently we have no construction of posets for which the boolean dimension achieves the logarithmic upper bound.

Proof of Proposition 1. Let $P_n = (X, <)$ be a fixed poset, $X = \{x_1, \dots, x_n\}$ Let $l_1, \dots, l_d$ be labellings $X \to \mathbf{N}$. The type $t(x_i, x_j)$, $i < j$, is the sequence $(e_l, \dots, e_d)$ defined by

$$e_k = \begin{cases} 1 & \text{if } l_k(x_i) < l_k(x_j), \\ -1 & \text{if } l_k(x_i) > l_k(x_j). \end{cases}$$

The set $\{l_1, \dots, l_d\}$ of labellings is called *distinguishing* if all pairs get different types (i.e. $t(x_i, x_j) \ne t(x_{i'}, x_{j'})$ whenever $(i, j) \ne (i', j')$). It is easy to see that a distinguishing family may be transformed to a formula $\varphi = \varphi(a_l, \dots, a_d)$ which proves $\dim_B(P) \le d$.

Explicitly, we may put $\varphi = \bigvee_{x<y} \varphi(x, y)$ where $\varphi(x, y) = b_1 \wedge b_2 \wedge \dots \wedge b_d$ with

$$b_i = \begin{cases} a_i & \text{if the } i\text{-th entry of } t(x, y) \text{ is } +1, \\ \neg a_i & \text{if the } i\text{-th entry of } t(x, y) \text{ is -1}. \end{cases}$$

Thus it suffices to find a distinguishing family of labellings. To do so we use a particular case of a result of Friedman (2): There exists a constant c such that for every set X of size n one can construct functions $f_1, \dots, f_d$, $d \le c \log n$, $f_i : X \to \{1, 2, 3, 4\}$ with the following property: if x, y, z, t are points of X and $g : \{x, y, z, t\} \to \{1, 2, 3, 4\}$ is a 1–1 mapping then $g = f_i \mid \{x, y, z, t\}$ for some $i \le d$. Now let labelling l_i be a monotone extension of f_i (i.e. $l_i(x) < l_i(y)$ providing $f_i(x) < f_i(y)$). One may check that the family $\{l_1, \dots, l_d\}$ is distinguishing. □

Proof of Proposition 2. d labellings on a set $X = \{x_1, \dots, x_n\}$ may produce no more than

$$d \cdot n! \cdot 2^{2^d}$$

posets. As there are at least $2^{n^2/4}$ posets on X we obtain easily $d \ge c' \log n$. □

Problems

A poset P is called planar if its Hasse-diagram is a planar graph. Although planar posets with either 0 or 1 have dimension ≤ 3, see [4], there are examples of planar posets with arbitrarily high dimension [3]. However the example given in [3] has boolean dimension ≤ 4. This leads to the following:

Problem 3.1. Is the boolean dimension of planar posets unbounded ?

An approach to solve (positively) the above problem is to prove a Ramsey–type statement.

Denote by $(\binom{P}{2})$ the set of all pairs of comparable elements of P; i.e.

$$\binom{P}{2} = \{\{x, y\} \mid x < y\}.$$

We say that a poset Q is 2-Ramsey for P if for every partition $\binom{Q}{2} = a_1 \cup a_2$ there exists a subposet P' of Q, $P' \simeq P$, such that $\binom{P}{2}' \subseteq a_i$.

Problem 3.2. Is it true that for every planar poset P there exists a 2-Ramsey planar poset Q?

It is a routine to derive a positive answer to 3.1 from a positive answer to 3.2. Note also, that for general posets 3.2 has been proved in [5].

References

[1] J. Friedman: Constructing $0(n \log n)$ size monotone formulae for the k-th elementary symmetric polynomial of n variables, *Proc. 25 th Annual Symp. on Foundations of Computer Science, IEEE* (1984), 506-515.

[2] G. Gambosi, J. Nešetřil, M. Talamo: Locally presented posets and their applications *(submitted to TCS)*.

[3] D. Kelly: On the dimension of partially ordered sets, *Discrete Math.* **35** (1981), 135-156.

[4] D. Kelly, W. T. Trotter: Dimension theory for ordered sets, *In: Ordered sets (ed. I.Rival), P. Reidel Publ. Co* (1982), 171-212.

[5] J. Nešetřil, V. Rödl: Combinatorial partitions of finite posets and lattices — Ramsey lattices, *Algebra Univ.* **19** (1984), 106-119.

[6] W. T. Trotter, J. I. Moore: The dimension of planar posets, *J. Comb. Th. (B)*, **22** (1977), 54-67.

J.Nešetřil
KAM MFF UK
Charles University
Malostranské nám. 25
118 00 Praha 1

P.Pudlák
MÚ ČSAV
Czechoslovak Acad. Sci.
Žitná 25
115 67 Praha 1

13. Intersection Properties and Extremal Problems for Set Systems

Zs. Tuza *

Abstract

A general inequality is proved for collections of finite sets satisfying a certain type of intersection properties. From this result, Ramsey–type theorems are deduced. For example, if every vertex is incident to edges of at most k colors in an edge–coloring of a complete graph on $2^k + 1$ vertices, then there is a monochromatic odd cycle.

1. Introduction

The primary motivation of this paper has been to point out a relation between two coloring concepts of graphs. As usual, denote by $\chi(G)$ the chromatic number of a graph $G = (V, E)$, i.e., $\chi(G)$ is the least number of classes in a partition $V_1 \cup \ldots \cup V_t = V$ in which each V_i is an independent vertex set of G. The other notion, introduced in [9], is called *local k–coloring* and means a coloring of the *edges* of G in such a way that each vertex is incident to edges of at most k colors. It should be stressed that there is no restriction on the number of colors used in a local k–coloring of G.

It is an easy exercise to show that if G is the union of k graphs of chromatic number $\leq t$ (i.e., $E = E_1 \cup \ldots \cup E_k$ and $\chi(G_i) \leq t$ for all $G_i = (V, E_i)$, $1 \leq i \leq t$) then $\chi(G) \leq t^k$. This property, however, does not hold for *local k*–colorings: As shown in [8], there exist graphs with arbitrarily large (and also with infinite) chromatic number and having a local 2–coloring in which all monochromatic subgraphs are bipartite. This observation indicates an essential difference between local k–colorings and "global k–colorings" (i.e., colorings by at most k colors).

* Research supported in part by the "AKA" Research Fund of the Hungarian Academy of Sciences

From this point of view it might be surprising that for local colorings of the complete graph K_n on n vertices the following result, analogous with the case of global colorings, can be proved (Theorem 5): If in a local k–coloring of K_n all monochromatic subgraphs have chromatic number at most t then $n \le t^k$.

A similar result can be formulated for directed paths of locally k–colored tournaments (directed graphs in which any two vertices are adjacent by precisely one arc). In this way, our Theorem 7 generalizes results, due to several authors (Rado [12], Gyárfás and Lehel [7], Chvátal and Komlós [4], Cvátal [3], Bermond [2]) concerning arc–colored tournaments.

Because of the above–mentioned difference, the standard techniques of treating global colorings cannot be applied to local colorings. This is the reason why we need more general structures, namely collections of finite sets satisfying some intersection properties. More precisely, t set systems $\mathcal{A}_i = \{A_{1,i}, \ldots, A_{n,i}\}$ $(i = 1, \ldots, t)$, consisting of the same number of sets, are considered, where any two sets having the same first subscript must be disjoint. Moreover, for pairs of distinct first subscripts some type of non–empty intersection is required. Such an assumption can be defined in various ways — for our purpose the most convenient one is formulated and investigated in Section 2. The main result of this type, Theorem 1, is a generalization of that in [15] and even its corollaries are strong enough to yield a sharp theorem for local colorings.

Some other possible sorts of intersection are indicated in Section 4, where an inequality related to so–called qualitatively independent partitions is proved. This result is not sharp, however, and there are several other related problems which should be the subject of future research. At first sight one might guess those intersection properties are useless and have some interest only in themselves. But this is not the case for $t = 2$: set–pair systems have found a great number of applications in various fields of extremal combinatorics. A detailed survey of them is given in [18]. For $t > 2$, the present work provides some examples how collections of finite sets can be used for handling graph theoretic problems, and the author hopes this machinery will prove to be useful in answering further questions in the future.

2. Collections of Finite Sets

Let $\mathcal{A}_j = \{A_{1,j}, A_{2,j}, \ldots, A_{n,j}\}$ $(j = 1, 2, \ldots, t)$ be t collections of finite sets with the following properties.

(1) $A_{i,j} \cap A_{i,l} = \emptyset$ for all i, j and l, $1 \le i \le n$, $1 \le j < l \le t$.

(2) For all i and j $(1 \leq i < j \leq n)$ there exist distinct l and s $(1 \leq l, s \leq t)$ such that $A_{i,l} \cap A_{j,s} \neq \emptyset$.

For such set systems we prove the following inequality.

Theorem 1. *Let $p_1, \ldots, p_t$ be arbitrary positive real numbers such that $p_1 + \ldots + p_t = 1$. Suppose that the set systems $\mathcal{A}_1, \ldots, \mathcal{A}_t$ satisfy (1) and (2). Then*

$$\sum_{i=1}^{n} \prod_{j=1}^{t} p_j^{|A_{i,j}|} \leq 1.$$

Proof. Let $X = \{v_1, \ldots, v_m\} = \cup_{i,j} A_{i,j}$. Take a random t–partition $\mathcal{P} = (P_1, \ldots, P_t)$ of X, where $P_1 \cup \ldots \cup P_t = X$ and the probability $\Pr(v_l \in P_j) = p_j$ for all l, $1 \leq l \leq m$. Suppose further that this probability distribution is independent of that belonging to all other v_k.

Denote by $\mathcal{E}_i$ the event that $A_{i,j} \subset P_j$ holds for every j, $1 \leq j \leq t$. Clearly,

$$\Pr(\mathcal{E}_i) = \prod_{j=1}^{t} p_j^{|A_{i,j}|}$$

for all i, $1 \leq i \leq n$. On the other hand, $\mathcal{E}_i$ and $\mathcal{E}_j$ cannot hold at the same time whenever $i \neq j$. Indeed, $\mathcal{E}_i$ implies $A_{i,l} \subset P_l$, while $\mathcal{E}_j$ yields $A_{j,s} \subset P_s$. By (2) and the definition of $\mathcal{E}_i$, $\emptyset \neq A_{i,l} \cap A_{j,s} \subset P_l \cap P_s$, contradicting the fact that $\mathcal{P}$ is a partition of X. Thus,

$$\Pr(\mathcal{E}_1) + \ldots + \Pr(\mathcal{E}_n) \leq 1.$$

□

It is worth formulating some particular cases of Theorem 1 separately, which will be used in applications. The first one is obtained by putting $p_1 = \ldots = p_t = \frac{1}{t}$.

Corollary 2. *If the set systems $\mathcal{A}_1, \ldots, \mathcal{A}_t$ satisfy (1) and (2) then*

$$\sum_{i=1}^{n} t^{-(\sum_{j=1}^{t} |A_{i,j}|)} \leq 1.$$

□

Corollary 3. *Suppose that the set systems $\mathcal{A}_1, \ldots, \mathcal{A}_t$ satisfy (1) and (2) and, for all i $(1 \leq i \leq n)$, $|A_{i,1}| + \ldots + |A_{i,t}| \leq k$. Then $n \leq t^k$.* □

Corollary 4. *Let $a_1, \ldots, a_t$ be given positive integers. Suppose that $\mathcal{A}_1, \ldots, \mathcal{A}_t$*

satisfy (1) and (2) and $|A_{i,j}| \le a_j$ *for all values of* i *and* j. *Then*

$$n \le \frac{(\sum_{j=1}^{t} a_j)^{\sum_{j=1}^{t} a_j}}{\prod_{j=1}^{t} a_j^{a_j}}.$$

Proof. Put $p_j = a_j/(a_1 + \ldots + a_t)$ in Theorem 1. □

Corollary 3 is sharp: Take a k–element underlying set Y and consider all partitions of Y into exactly t classes numbered from 1 to t, where a partition class is allowed to be empty. Define $A_{i,j}$ as the jth class of the ith partition. Then (1) holds by the definition of a partition and (2) is implied by the fact that $A_{i,1} \cup \ldots \cup A_{i,t} = Y$ for every i. Moreover, the number of t–partitions is equal to t^k, the number of functions $Y \to \{1, \ldots, t\}$.

On the other hand, Corollary 4 is not sharp, even in the particular case $t = 2$. It would be interesting to determine the maximum value of n, under the assumptions of Corollary 4. Some related results for $t = 2$ are given in [15].

3. Applications

In this section we apply the corollaries of Theorem 1 in the solution of some graph theoretic problems. First we deal with Ramsey–type theorems and then consider coverings of K_n.

3.1. Local Ramsey Theorems

It is easy to color the edges of the complete graph on t^k vertices by k colors in such a way that each monochromatic subgraph has chromatic number $\le t$. Indeed, let $V(K_n) = \{\mathbf{x} : \mathbf{x} = (\mathbf{x}_1, \ldots, \mathbf{x}_k),\ \mathbf{x}_i \in \{1, \ldots, t\}\}$ and define the color of the edge $(\mathbf{x}, \mathbf{x}')$ to be j if and only if the first $j-1$ coordinates of $\mathbf{x}$ and $\mathbf{x}'$ are identical but they differ in the jth coordinate. Since every global k–coloring is also a local k–coloring, this example shows that the following result is sharp.

Theorem 5. *In every local k–coloring of the complete graph on* $n \ge t^k + 1$ *vertices there exists a monochromatic subgraph whose vertex chromatic number is larger than* t.

Proof. Suppose that $E(K_n) = E(G_1) \cup \ldots \cup E(G_m)$, where each $v_i \in V = \{v_1, \ldots, v_n\}$ is contained in at most k graphs G_j, and let $\chi(G_j) \leq t$ for all j, $1 \leq j \leq m$. Moreover, denote by $V_{j,l}$ the lth color class in an arbitrary fixed t–coloring of G_j. For $i = 1, \ldots, n$ and $l = 1, \ldots, t$, set $A_{i,l} = \{j : v_i \in V_{j,l}\}$.

If $v_i \in V_{q,j}$ (for some $q \leq m$ and $j \leq t$) then $v_i \notin V_{q,l}$ for any $l \neq j$, i.e., $A_{i,j} \cap A_{i,l} = \emptyset$ for all i, j and l $(j \neq l)$. On the other hand, since every edge v_iv_j of K_n is contained in some G_q, those two vertices belong to different classes of G_q; say, $v_i \in V_{q,l}$ and $v_j \in V_{q,s}$. Then $q \in A_{i,l} \cap A_{j,s} \neq \emptyset$. Thus, (1) and (2) are satisfied by the sets $A_{i,j}$, $1 \leq i \leq n$, $1 \leq j \leq t$. Observe further that $|A_{i,1}| + \ldots + |A_{i,t}| \leq k$ for all i, since the left-hand side is equal to the number of the G_j containing the vertex v_i, and we have a local k–coloring of K_n. Thus, $n \leq t^k$ by Corollary 3. □

Probably the most interesting particular case of this result is when $t = 2$:

Corollary 6. *In every local k–coloring of a complete graph on more than 2^k vertices there can be found a monochromatic odd circle.* □

It is a natural question how large n should be as a function of k and s if a monochromatic odd circle of given length $2s + 1$ is to be found in a locally k–colored K_n. For global colorings, Erdős and Graham [5] proved the lower and upper bounds of $s2^k$ and $2s(k+1)!$, respectively. In [17] we show that similar bounds hold for local colorings, i.e., the estimates are exponential functions of k. In contrast, if the monochromatic subgraph to be found is cycle–free (a tree or a forest) and has s edges then the estimate is a linear function of both k and s.

A well-known result of Gallai [6] and Roy [13] states that a graph has chromatic number at most t if and only if its edges can be oriented in such a way that no directed path of t edges occurs. Hence, as observed by the authors of [8], Theorem 5 implies the following result.

Theorem 7. *If the arcs of a tournament T on $n > t^k$ vertices are locally k–colored then there can be found a monochromatic directed path of at least t arcs in T.* □

As mentioned in the introduction, for global k–colorings Theorem 7 and its particular case $k = 2$ have been observed by several authors ([2], [3], [4], [7], [12]). Though, in a sense, Theorem 7 looks simpler than Theorem 5 (and perhaps the reader would expect an easy proof for it), in fact the two results are equal in strength.

3.2. Covering of K_n by Subgraphs of Low Chromatic Number

Here we prove a common generalization of Theorem 5 and a result of Katona and Szemerédi [10]. Its proof is based on Theorem 1 and a convexity argument which appears in [10] and also in [14].

Theorem 8. *Let $G_1, \ldots, G_m$ be subgraphs of K_n and suppose $E(G_1) \cup \ldots \cup E(G_m) = E(K_n)$. If $\chi(G_j) \leq t$ for all j, $1 \leq j \leq m$, then $|V(G_1)| + \ldots + |V(G_m)| \geq \frac{n \log n}{\log t}$.*

Proof. As in the proof of Theorem 5, we fix a t–coloring $V_{j,1}, \ldots, V_{j,t}$ of every G_j and set $A_{i,l} = \{j : v_i \in V_{j,l}\}$. These sets $A_{i,l}$ satisfy (1) and (2), so that Corollary 2 implies

$$\sum_{i=1}^{n} t^{-(\sum_{j=1}^{t} |A_{i,j}|)} \leq 1.$$

Since the function $f(x) = t^{-x}$ is convex, putting

$$\sigma = \sum_{i=1}^{n} \sum_{j=1}^{t} |A_{i,j}|,$$

we obtain

$$1 \geq \sum_{i=1}^{n} t^{-(\sum_{j=1}^{t} |A_{i,j}|)} \geq nt^{-\sigma/n},$$

thus, $\sigma \geq n \log n / \log t$. On the other hand,

$$\begin{aligned} |V(G_1)| + \ldots + |V(G_m)| &= \sum_{i=1}^{t} |\{j : v_i \in V(G_j)\}| = \\ &= \sum_{i=1}^{n} (|A_{i,1}| + \ldots + |A_{i,t}|) = \sigma \end{aligned}$$

and the theorem follows. □

Note that the above result is valid for every $t \geq 2$, i.e., t can tend to infinity as $n \to \infty$. This fact will be used in a proof of a theorem concerning decompositions of graphs into perfect subgraphs in [16].

4. Further Types of Prescribed Intersections

For set systems considered in Section 2, the case $t = 2$ can be interpreted in the following way. We have n pairs (A_i, B_i) of finite sets and assume that

$A_i \cap B_i = \emptyset$ for all i. Then (2) means that $A_i \cap B_j$ or $A_j \cap B_i$ is nonempty whenever $i \neq j$. Even in this particular case, however, there are two further intersection properties which are reasonable to assume and which have also gained several applications. Namely, instead of (2) one can assume $A_i \cap B_j \neq \emptyset$ for all $i < j$ or, as an even stronger requirement, for all $i \neq j$. Moreover, Alon [1] considers another variant, assuming that a partition is given on the underlying set and the cardinality of the sets is bounded within each partition class.

Hence, even the simplest particular case provides a considerable number of questions, and this number increases exponentially as t gets large. Those properties like (2) can be defined by bipartite graphs $B = (V_1, V_2; E)$ where $V_1 \cap V_2 = \emptyset$, $V_1 = \{x_1, \ldots, x_t\}$, $V_2 = \{y_1, \ldots, y_t\}$, and each $e \in E$ joins some x_l to some y_s. Now the set systems $\mathcal{A}_1, \ldots, \mathcal{A}_t$ can be viewed as *one* collection $\mathcal{A} = \{(A_{i,1}, \ldots, A_{i,t}) : 1 \leq i \leq n\}$ of t–tuples of finite sets, and we say the pair i, j $(1 \leq i < j \leq n)$ is B–intersecting if the set $\{(l, s) : A_{i,l} \cap A_{j,s} \neq \emptyset\}$ of intersections contains $\{(l, s) : x_l y_s \in E\}$. Suppose a family $\mathcal{B} = \{B_1, \ldots, B_r\}$ of bipartite graphs is given. Call $\mathcal{A}$ a *$\mathcal{B}$–intersecting collection* if it satisfies (1) and each possible pair of integers i and j is B_k–intersecting for some $k \leq r$. In this way, (2) corresponds to the case when $\mathcal{B}$ consists of all one–edge graphs $B_{l,s} = (V_1, V_2; \{x_l y_s\})$, $l \neq s$, $1 \leq l, s \leq t$.The natural generalization of the above mentioned properties for $t = 2$ is $\mathcal{B}' = \{B_{l,s} : 1 \leq l < s \leq t\}$ and $\mathcal{B}'' = \{K_{t,t} \setminus tK_2\}$ where $K_{t,t} \setminus tK_2$ denotes the complete bipartite graph from which the t pairwise disjoint edges $x_1y_1, \ldots, x_ty_t$ have been deleted. Also, the graphs with edge set $\{x_i y_j : 1 \leq i < j \leq t\}$ or $\{x_i y_j : 1 \leq i \leq j \leq t\}$ or having a given number of edges, seem to be interesting.

In the present paper we do not intend to study those properties in detail but we mention one application of $\mathcal{B}''$–intersecting collections. Let $\mathcal{P} = \{P_1, \ldots, P_n\}$ be a family of partitions $P_i = \{P_{i,1}, \ldots, P_{i,t}\}$ of a given underlying set X into t classes. Then $\mathcal{P}$ is called *r–wise qualitatively independent* (r–QI, for short) if for all distinct $i_1, \ldots, i_r$ and for all $j_1, \ldots, j_r$, $P_{i_1 j_1} \cap \ldots \cap P_{i_r j_r} \neq \emptyset$ holds. An interesting question, arising in probability theory and also having relation to the theory of Boolean functions and error correcting codes, is to determine the maximum number $m(r, t, n)$ of r–QI partitions over an n–element underlying set. This general question can reasonably be reduced to the case $r = 2$, and then any 2-QI family is a $K_{t,t}$–intersecting collection. So far, the best upper bound for $m(2, t, n)$ has been achieved by Poljak and the author [11], using the idea that from any $K_{t,t}$–intersecting collection there can be obtained a suitable $K_{2,2}$–intersecting one. On the other hand, somewhat surprisingly, the lower bounds for $m(2, t, n)$ could be improved by a construction using $\mathcal{B}''$–intersecting collections.

Below we use the "permutation method" for proving two inequalities. The first one can be viewed as a symmetric version of Theorem 1, since its supposition is a variant of (2); the second one considers a stronger intersection property and is related to the problem of QI–partitions.

Proposition 9. *Suppose that the set system* $\mathcal{A} = \{A_{i,j} : 1 \le i \le n,\ 1 \le j \le t\}$ *satisfies (1) and*
(2') for all i *and* j $(1 \le i < j \le n)$ *there exist distinct* l *and* s $(1 \le l, s \le t)$ *such that* $A_{i,l} \cap A_{j,s} \ne \emptyset$ *and* $A_{i,s} \cap A_{j,l} \ne \emptyset$.

Then

$$\sum_{i=1}^{n}((\prod_{j=1}^{t}|A_{i,j}|!)/(\prod_{j=1}^{t}|A_{i,j}|)!) \le 1.$$

Proof. Set $X = \cup_{i,j}A_{i,j}$ and suppose $X = \{1,2,\ldots,m\}$. Let $\mathcal{P}$ be the collection of all permutations of X, $|\mathcal{P}| = m!$. For every i, $1 \le i \le n$, define

$$\mathcal{P}_i = \{\pi \in \mathcal{P} : \pi(A_{i,l}) < \pi(A_{i,s}) \text{ for } 1 \le l < s \le t\}$$

where $\pi(A) < \pi(A')$ means "$\pi(x) < \pi(x')$ for every $x \in A$, $x' \in A'$." Observe that $\mathcal{P}_i \cap \mathcal{P}_j = \emptyset$ whenever $i \ne j$. Indeed, $\pi \in \mathcal{P}_i \cap \mathcal{P}_j$ would imply $\pi(A_{i,l}) < \pi(A_{i,s})$ and $\pi(A_{j,l}) < \pi(A_{j,s})$ for all $l < s$, so that (2') cannot hold. Moreover

$$|\mathcal{P}_i| = m!(\prod_{j=1}^{t}|A_{i,j}|!)/(\sum_{j=1}^{t}|A_{i,j}|)!.$$

Thus,

$$m! = |\mathcal{P}| \ge |\mathcal{P}_1 \cup \ldots \cup \mathcal{P}_n| = |\mathcal{P}_1| + \ldots + |\mathcal{P}_n| =$$
$$= m!\sum_{i=1}^{n}((\prod_{j=1}^{t}|A_{i,j}|!)/(\sum_{j=1}^{t}|A_{i,j}|)!)$$

and the statement follows. □

Proposition 10. *Let* $\mathcal{A} = \{A_{i,1},\ldots,A_{i,t} : 1 \le i \le n\}$ *be a* $\mathcal{B}''$*-intersecting collection. Then*

$$\sum_{i=1}^{n}\Bigg(\sum_{1\le j_1<j_2\le t}\frac{|A_{i,j_1}|!|A_{i,j_2}|!}{(|A_{i,j_1}|+|A_{i,j_2}|)!} - \sum_{1\le j_1<j_2<j_3\le t}\frac{|A_{i,j_1}|!(|A_{i,j_2}|+|A_{i,j_3}|)!+|A_{i,j_3}|!(|A_{i,j_1}|+|A_{i,j_2}|)!}{(|A_{i,j_1}|+|A_{i,j_2}|+|A_{i,j_3}|)!}\Bigg) \le 1$$

Proof. As in the previous proof, let $X = \cup_{i,j}A_{i,j} = \{1,\ldots,m\}$ and let $\mathcal{P}$ be the set of all permutations of X. Set

$$\mathcal{P}_i = \{\pi \in \mathcal{P} : \pi(A_{i,j_1}) < \pi(A_{i,j_2}) \text{ for some } j_1 \text{ and } j_2,\ 1 \le j_1 < j_2 \le t\}.$$

Since $\mathcal{A}$ is $\mathcal{B}''$–intersecting, we obtain $\mathcal{P}_i \cap \mathcal{P}_j = \emptyset$ whenever $i \neq j$. Now the statement follows by the "sieve" principle: The first term in the inequality stands for the number of triplets (π, j_1, j_2) such that $\pi(A_{i,j_1}) < \pi(A_{i,j_2})$ while the second term is an upper bound for triplets counted at least twice. (If j_1, j_2, j_3, j_4 are four distinct subscripts such that $\pi(A_{i,j_1}) < \pi(A_{i,j_2})$ and $\pi(A_{i,j_3}) < \pi(A_{i,j_4})$ then $\max \pi(A_{i,j_1}) < \max \pi(A_{i,j_3})$ implies $\pi(A_{i,j_1}) < \pi(A_{i,j_4})$, so that we need not subtract any term for 4–tuples of subscripts.) □

The reason why Proposition 10 can be used for improving estimates on the number of QI–partitions is that if every class $A_{i,j}$ has nearly the same cardinality then the second term is rather too small as $|X| \to \infty$; on the other hand, if some $A_{i,j}$ is much smaller than the average then the terms involving $A_{i,j}$ are much larger than the other ones, i.e., in this latter case an immediate improvement is obtained. We do not enter into details, however, since the gain is a factor of $(1 + o(1))t^2$ only, while the gap between current upper and lower bounds is an exponential function of the cardinality of the underlying set (cf. [11]).

Acknowledgements. I am grateful to J. Lehel for fruitful discussions concerning Section 3, to A. Gyárfás for his helpful suggestions, and to P. Frankl for his remarks on [15] that were useful when preparing the present paper also.

Notes added in November 1987

1. As J. Körner kindly informed us, an alternative proof of our Theorem 8 can be derived from his results on graph entropy, namely Lemmas 1, 2 and K^* of the paper [19].

2. Recently, some generalizations of Theorem 1 have been found by Y. Caro and the present author. By the new inequalities, Theorems 5 and 8 can be extended for r–uniform hypergraphs also. Details will be given in a forthcoming joint paper.

References

[1] N. Alon, An extremal problem for sets with applications to graph theory, *J. Combinatorial Theory, Ser. A*, **40** (1985) 82-89.

[2] J.C. Bermond, Some Ramsey numbers for directed graphs, *Discrete Math.* **9** (1974) 313-321.

[3] V. Chvátal, Monochromatic paths in edge–colored graphs, *J. Combinatorial Theory, Ser. B.*, **13** (1972) 69-70.

[4] V. Chvátal and J. Komlós, Some combinatorial theorems on monotonicity, *Canad. Math. Bull.* **14** (1971) 151-157.

[5] P. Erdős and R.L. Graham, On partition theorems for finite graphs, in *Infinite and Finite Sets*, Coll. Math. Soc. János Bolyai **10** (Keszthely 1973), 515-527.

[6] T. Gallai, On directed paths and circuits, in *Theory of Graphs (P. Erdős and G.O.H. Katona, eds)*, Academic Press, 1968, 115-118.

[7] A. Gyárfás and J. Lehel, A Ramsey–type problem in directed and bipartite graphs, *Periodica Math. Hungar.* **3** (1973) 299-304.

[8] A. Gyárfás, J. Lehel, J. Nešetřil, V.Rödl, R.H. Schelp and Zs. Tuza, Local k–colorings of graphs and hypergraphs, *J. Combinatorial Theory, Ser. B,* **43** (1987) 127-139.

[9] A. Gyárfás, J. Lehel, R.H. Schelp and Zs. Tuza, Ramsey numbers for local colorings, *Graphs and Combinatorics*, **3** (1987) 267-277.

[10] G.O.H. Katona and E. Szemerédi, On a problem of graph theory, *Studia Sci. Math. Hungar.* **2** (1987) 23-28.

[11] S. Poljak and Zs. Tuza, On the maximum number of qualitatively independent partitions, *submitted.*

[12] R. Rado, Theorems on the colouring of the edges of a graph, in:*Combinatorial Mathematics and its Applications (C. Bose et al., eds.)*, Proc. 2nd Chapel Hill Conf., Univ. North Carolina, 1970, 385-390.

[13] R. Roy, Nombre chromatique et plus longs chemins d'un graphe, *Revue AFIRO* **1** (1967) 127-132.

[14] T.G. Tarján, Complexity of lattice–configurations, *Studia Sci. Math. Hungar.* **10** (1975) 203-211.

[15] Zs. Tuza, Inequalities for two set systems with prescribed intersections, *Graphs and Combinatorics* **3** (1987) 75-80.

[16] Zs. Tuza, Perfect graph decompositions, *in preparation.*

[17] Zs. Tuza, Ramsey numbers for forests and sparse graphs, *in preparation.*

[18] Zs. Tuza, Applications of set pairs in extremal combinatorial problems, *in preparation.*

[19] J. Körner, Fredman–Komlós bounds and information theory, *Siam J. Alg. Discr. Methods* **7** (1986) 560-571.

Zs. Tuza
Computer and Automation Institute
Hungarian Academy of Sciences
H–1111 Budapest, Kende u. 13–17
Hungary

14. On an Imbalance Problem in the Theory of Point Distribution

G. Wagner

Abstract

We consider the following problem: Let f be a 2π-periodic integrable function satisfying $\int_0^{2\pi} f(x)dx=0$. Given an N-tuple of points $\omega_N=\{x_1,x_2,\dots,x_N\}$ on $[0,2\pi)$, denote by $\mathrm{Pos}(f,\omega_N)$ the set of all $x\in[0,2\pi)$ for which $\sum_{j=1}^N f(x-x_j)\geq 0$ is true. Let $\beta_N(f)=\inf m(\mathrm{Pos}(f,\omega_N))$, where m denotes Lebesgue measure on $[0,2\pi)$ and the infimum is taken over all N-tuples ω_N.

We give lower and upper bounds for $\beta_N(f)$ in three special cases, together with some results of a more general type.

1. Introduction

Let $f(x)$ be a real valued, 2π-periodic function on $\mathbf{R}$, Lebesgue-integrable on $[0,2\pi)$ such that $\int_0^{2\pi} f(x)dx = 0$ holds. For some finite point set $\omega_N = \{x_1,x_2,\dots,x_N\} \subset [0,2\pi)$ consider the sum of translates $\sum_{j=1}^N f(x-x_j)$. We are interested in the Lebesgue measure m of the set of points $x \in [0,2\pi)$ for which $\sum_{j=1}^N f(x-x_j) \geq 0$ holds.

Problem A. Can we find, for each $\varepsilon > 0$, a set $\{x_1,x_2,\dots,x_{N(\varepsilon)}\}$ such that $m\{x\in[0,2\pi) : \sum_{j=1}^N f(x-x_j)\geq 0\} < \varepsilon$ is true? Let $f(x) \sim \sum_{n=1}^\infty (a_n \cos nx + b_n \sin nx)$ be the Fourier expansion of $f(x)$, and let $H \subset \mathbf{N}$ be the subset of indices n with $a_n^2+b_n^2>0$. Problem A is equivalent to the following

Problem A'. Given a subset $H \subset \mathbf{N}$, does there exist, for each $\varepsilon > 0$, a trigonometric polynomial $\sum_{h\in H}(a_h \cos hx + b_h \sin hx)$ such that $m\{x\in[0,2\pi) : \sum_{h\in H}(a_h\cos hx + b_h \sin hx)\geq 0\} < \varepsilon$ holds?

The answer to Problem A'is certainly positive if $H = \mathbf{N}$ or $H = a\mathbf{N}$ for some $a \in \mathbf{N}$. In this case we may choose the Fejér kernel $\sum_{h=1}^N (N-h+1)\cos hax$ with $N = N(\varepsilon)$ sufficiently large. The answer is negative if $H \cap b\mathbf{N}$

is finite for some $b \in \mathbf{N}$. Another criterion of negative character is given below. We may ask

Problem B. In dependence of N, give lower and upper bounds for the infimum of the measure $m\{x \in [0, 2\pi) : \sum_{j=1}^{N} f(x - x_j) \geq 0\}$, the infimum being taken over all N-tuples $\{x_1, x_2, \ldots, x_N\}$.

In the sequel, we shall primarily be concerned with the discussion of Problem B for some interesting special cases. However, Problems A and B in their general form show some aspects that seem worth mentioning, and we shall list some of the more general results, omitting the proofs.

2. Some General Results

Let $f(x) \in L^1_{\text{per}}[0, 2\pi)$. Denote by Pos$f$ the set of points $x \in [0, 2\pi)$ for which $f(x) \geq 0$ holds. For some N-tuple $\omega_N = \{x_1, \ldots, x_N\}$ denote by Pos(f, ω_N) the set of points $x \in [0, 2\pi)$ for which the inequality $\sum_{j=1}^{N} f(x - x_j) \geq 0$ is true. Finally, let $\beta_N(f) = \inf m(\text{Pos}(f, \omega_N))$, the infimum being taken over all N-tuples $\omega_N = \{x_1, x_2, \ldots, x_N\}$.

(a) Assume that $(\omega_N) = (\{x_1^{(N)}, x_2^{(N)}, \ldots, x_N^{(N)}\})$ is a sequence of N-tuples such that $\lim_{N\to\infty} m(\text{Pos}(f, \omega_N)) = 0$ holds. Assume further, that $f(x) \sim \sum_{n=1}^{\infty} (a_n \cos nx + b_n \sin nx)$ with $a_n^2 + b_n^2 > 0$ for each n. Then, as N tends to infinity, the distribution of the N-tuples ω_N converges weakly to uniform distribution.

(b) for any $f(x) \in L^p_{\text{per}}[0, 2\pi) (1 < p \leq \infty, \frac{1}{p} + \frac{1}{q} = 1)$ the inequality $m(\text{Pos}f) \geq (\| f \|_1 / 2 \| f \|_p)^q$ holds, where $\| f \|_p^p = \frac{1}{2\pi} \int_0^{2\pi} | f(x) |^p \, dx$. This elementary inequality shows, for example, that if $f(x) \in L^2[0, 2\pi)$ and $f(x) \sim \sum_{j=1}^{\infty} (a_{h_j} \cos h_j x + b_{h_j} \sin h_j x)$ with $h_{j+1}/h_j \geq \lambda > 1$ for all j, then the answer to problem A is negative. Namely we have $\| \sum_{j=1}^{N} f(x - x_j) \|_2 \leq c \cdot \| \sum_{j=1}^{N} f(x - x_j) \|_1$, with $c > 0$ depending only on the lacunary sequence (h_j) (see f.e. [3], p.108).

(c) Let $f(x) \sim \sum_{h=1}^{\infty} (a_h \cos hx + b_h \sin hx)$ with $a_h^2 + b_h^2 > 0$ for all values of h. Assume that $f(x)$ is of bounded variation $V(f)$ on $[0, 2\pi)$, and let n be a natural number. Then for each $N > V(f) \cdot \sqrt{n} \sum_{h=1}^{n} (a_h^2 + b_h^2)^{-1/2}$ there exists a set of points $\omega_N = \{x_1, x_2, \ldots, x_N\}$ such that the inequality $m(\text{Pos}(f, \omega_N)) < 4\pi/\sqrt{n}$ holds.

The proof of this latter result uses the following simple idea. Convolution of $f(x)$ with an appropriate positive trigonometric polynomial yields the Fejér kernel $\sum_{k=1}^{n} (n - k + 1) \cos kx$ which is greater than $-\frac{1}{2}\sqrt{n+1}$ on a set of measure less than $4\pi/\sqrt{n}$. Now approximate this convolution by a convolution with a discrete measure, and use Koksma's inequality for numerical integration.

In the following examples we use special methods to give lower and upper bounds for the quantities $\beta_N(f)$.

3. On the Product of Distances with Respect to a Point Set

We wish to distribute N points $z_1, z_2, \ldots, z_N$ on the unit circle in such a way that the linear measure of the set of points $z, |z| = 1$, for which the inequality $\sum_{j=1}^{N} |z - z_j| \leq 1$ holds, becomes minimal. This is Problem B for the function $f(x) = -\log|2\sin\frac{x}{2}|$. Erdős, Herzog and Piranian [1] suggested a distribution of the points z_k which gives an upper bound $\beta_N(f) \leq 16(\frac{\log N}{N})^{1/3}$ for $N \geq 2$ (see [5]).

Using the energy inequality $\sum_{1\leq j<k\leq r} \alpha_j\alpha_k \log |z_j - z_k| \leq \sum_{j=1}^{r} \alpha_j^2 \log\frac{N\cdot e^2}{\alpha_j}$, valid for any r-tuple of points $z_1, z_2, \ldots, z_r$ on the unit circle, and any set of positive integers $\alpha_1, \alpha_2, \ldots, \alpha_r$ satisfying $\alpha_1 + \alpha_2 + \ldots + \alpha_r = N$, the author [5] proved the lower bound $\beta_N(f) \geq 1/4\sqrt{N}$ for all N. The present linear problem has an interesting counterpart in the complex plane. Consider N-tuples $\omega_N = \{z_1, z_2, \ldots, z_N\}$ in the unit disk $|z| \leq 1$. Denote by $A(\omega_N)$ the area of the lemniscate domain $\{z : \sum_{k=1}^{N} |z - z_k| \leq 1\}$. Using a theorem due to G. MacLane, the authors in [1] proved the existence of N-tuples ω_N even on the unit circle with arbitrarily small $A(\omega_N)$, provided that N is large enough. The principal argument used there depends on Runge's theorem and is nonconstructive, and no nontrivial upper or lower bounds for $\inf_{\omega_N} A(\omega_N)$ are hitherto known.

4. On the Sum of Distances with Respect to a Point Set

We wish to distribute N points $z_1, z_2, \ldots, z_N$ on the unit circle in such a way that the measure of the set of points $z, |z| = 1$, for which the inequality $\sum_{j=1}^{N} |z - z_j| \leq N \cdot \frac{4}{\pi}$ holds, becomes minimal. This is Problem B with $f(x) = (4/\pi) - |2\sin\frac{1}{2}x|$.

To get a lower estimate for $\beta_N(f)$, the method developed in [5] can be used once a good upper estimate for the energy sum $\sum_{1\leq j<k\leq r} \alpha_j\alpha_k |z_j - z_k|$, $\sum_{j=1}^{r} \alpha_j = N$, $\alpha_j \geq 1$, is known.

In the case when all the α_j are equal to 1, it was proved by Fejes Tóth [2] that $\sum_{1\le j<k\le N} \mid z_j - z_k \mid$ becomes maximal is the z_k are the vertices of a regular N-gon. In this case we have

$$\sum_{j<k} \mid z_j - z_k \mid = N \cot \frac{\pi}{2N} = \frac{2N^2}{\pi} - \frac{\pi}{6} + 0(\frac{1}{N^2})$$

We prove

Lemma. *Let $z_1, z_2, \ldots, z_r$ be points on the unit circle, let $\alpha_1, \alpha_2, \ldots, \alpha_r$ be positive integers satisfying $\sum_{j=1}^r \alpha_j = N$. The following energy inequality holds:*

$$\sum_{j<k} \alpha_j \alpha_k \mid z_j - z_k \mid \le \frac{2N^2}{\pi} - \frac{\pi}{6N} \sum_{j=1}^r \alpha_j^3.$$

Proof. Using the Fourier expansion $\mid \sin \frac{x}{2} \mid = \frac{2}{\pi} - \frac{4}{\pi} \sum_{h=1}^\infty \frac{\cos hx}{4h^2-1}$, and writing $z_1, \ldots, z_r$ in the form $e^{ix_1}, \ldots, e^{ix_r}$, we get:

$$\sum_{j<k} \alpha_j \alpha_k \mid z_j - z_k \mid = \sum_{j,k} \alpha_j \alpha_k \mid \sin \frac{1}{2}(x_j - x_k) \mid =$$

$$= \frac{2}{\pi} \sum_{j,k} \alpha_j \alpha_k - \frac{4}{\pi} \sum_{h=1}^{\infty} \sum_{j,k} \alpha_j \alpha_k \frac{\cos h(x_j - x_k)}{4h^2 - 1} =$$

$$= \frac{2}{\pi} N^2 - \frac{4}{\pi} \sum_{h=1}^{\infty} \frac{1}{4h^2-1} \mid \sum_{j=1}^r \alpha_j e^{ihx_j} \mid^2 \le$$

$$\le \frac{2}{\pi} N^2 - \frac{1}{\pi} \sum_{h=1}^{\infty} \frac{1}{h^2} \mid \sum_{h=1}^r \alpha_j e^{ihx_j} \mid^2 .$$

Now let $g(x) = \frac{1}{2}(\pi - x)$ on $[0, 2\pi)$, periodically continued over the real axis. We have

$$\int_0^{2\pi} \mid \sum_{j=1}^r \alpha_j g(x - x_j) \mid^2 dx = \int_0^{2\pi} \sum_{h=1}^{\infty} \frac{\sin^2 hx}{h^2} \mid \sum_{j=1}^r \alpha_j e^{ihx_j} \mid^2 dx =$$

$$\pi \sum_{h=1}^{\infty} \frac{1}{h^2} \mid \sum_{h=1}^r \alpha_j e^{ihx_j} \mid^2,$$

hence

$$\text{(1)} \qquad \frac{2}{\pi} N^2 - \sum_{j<k} \alpha_j \alpha_k \mid z_j - z_k \mid \ge \frac{1}{\pi^2} \int_0^{2\pi} \mid \sum_{j=1}^r \alpha_j g(x - x_j) \mid^2 dx$$

The function $\sum_{j=1}^r \alpha_j g(x - x_j)$ is piecewise linear, with jump discontinuities of height $\pi\alpha_j$ in the points x_j, and constant slope $-\frac{1}{2}N$. This observation allows us to calculate the minimal value of the righthand side in (1), which is easily seen to be equal to

$$\frac{1}{\pi^2}\sum_{j=1}^{r} 2\int_0^{\pi\alpha_j/N}\left(-\frac{N}{2}x\right)^2 dx = \frac{\pi}{6N}\sum_{j=1}^{r}\alpha_j^3.$$

This proves the lemma. □

Theorem 1. *Let $\omega_N = \{z_1, z_2, \ldots, z_N\}$ be an N-tuple of points on the unit circle. Let $S(\omega_N) = \{|z| = 1 : \sum_{j=1}^{N} |z - z_j| \le \frac{4}{\pi}N\}$. The following inequality holds:*

$$N^{-2/3} \le \inf_{\omega_N} m(S(\omega_N)) \le 30 \cdot N^{-1/2}.$$

Proof. In order to obtain the lower bound, we proceed as in [5], using the convexity of the function $u(x) = |2\sin\frac{x}{2}|$. Hence, without loss of generality, we may assume that $S(\omega_N)$ consists of $r \le N$ disjoint closed arcs $A_1, A_2, \ldots, A_r$, each containing a single point $e^{ix_1}, \ldots, e^{ix_r}$ with multiplicity $\alpha_1, \ldots, \alpha_r$; $\sum_{j=1}^{r}\alpha_j = N$.

Denote by $e^{ix_1'}, e^{ix_1''}$ the boundary points of the arc A_1. Let A_1', A_1'' be the arcs connecting e^{ix_1} with $e^{ix_1'}$, $e^{ix_1''}$, respectively. We have $\frac{4}{\pi}\alpha_1 N = \alpha_1\sum_{k=1}^{r}\alpha_k u(x_k - x_1') = \alpha_1\sum_{k=1}^{r}\alpha_k u(x_k - x_1'')$. Noting that $\alpha_1 u(x_1 - x_1') = 2\alpha_1\sin\frac{1}{2}m(A_1')$ and $\alpha_1 u(x_1 - x_1'') = 2\alpha_1 \sin\frac{1}{2}m(A_1'')$, and using convexity of the function $\sum_{k=2}^{r}\alpha_k u(x - x_k)$ on A_1, we get:

$$\begin{aligned}
&\alpha_1\sum_{k=2}^{r}\alpha_k u(x_k - x_1) \ge \\
&\ge \frac{m(A_1'')}{m(A_1)}\alpha_1\sum_{k=2}^{r}\alpha_k u(x_k - x_1') + \frac{m(A_1')}{m(A_1)}\alpha_1\sum_{k=2}^{r}\alpha_k u(x_k - x_1'') = \\
&\frac{4}{\pi}N\alpha_1 - \alpha_1^2\frac{m(A_1'')}{m(A_1)}\cdot 2\sin\frac{m(A_1')}{2} - \alpha_1^2\frac{m(A_1')}{m(A_1)}\cdot 2\sin\frac{m(A_1'')}{2} \ge \\
&\frac{4}{\pi}N\alpha_1 - 2\alpha_1^2\sin\frac{1}{4}m(A_1) \ge \frac{4}{\pi}N\alpha_1 - \frac{1}{2}\alpha_1^2 m(A_1).
\end{aligned} \tag{2}$$

Repeating the argument for the components $A_2, A_3, \ldots, A_r$, and adding the inequalities which correspond to (2), we obtain:

$$\sum_{j,k}\alpha_j\alpha_k u(x_j - x_k) \ge \frac{4}{\pi}N^2 - \frac{1}{2}\sum_{j=1}^{r}\alpha_j^2 m(A_j).$$

Comparing this result with the Lemma we get the final inequality

$$\frac{4}{\pi}N^2 - \frac{\pi}{3N}\sum_{j=1}^{r}\alpha_j^3 \ge \frac{4}{\pi}N^2 - \frac{1}{2}\sum_{j=1}^{r}\alpha_j^2 m(A_j)$$

or

$$\sum_{j=1}^{r} \alpha_j^2 m(A_j) \geq \frac{2\pi}{3N} \sum_{j+1}^{r} \alpha_j^3.$$

Let $\lambda = \max \alpha_j$. We have $\lambda^2 \sum_{j=1}^{r} m(A_j) \geq \frac{2\pi}{3N} \sum_{j=1}^{r} \alpha_j^3 \geq \frac{2\pi}{3N}(\lambda^3 + N - \lambda)$, or $m(S(\omega_N)) \geq \frac{2\pi}{3N}(\lambda + \frac{N}{\lambda^2} - \frac{1}{\lambda})$.

For all values of λ, $1 \leq \lambda \leq N$, this latter expression is greater than $N^{-2/3}$. This proves the lower estimate.

To get the upper estimate, first consider the function $g_N(x) = \sum_{k=0}^{N-1} u(x - \frac{2\pi}{N}k)$. We have $g_N(x) \geq 2 \cot \frac{\pi}{2N} \geq \frac{4}{\pi}N - \frac{\pi}{2N}$ for all $x \in [0, 2\pi)$ and for all N. For some natural number h, $h < \frac{1}{2}N$, replace the points $\frac{2\pi}{N}k$ ($k = 0, \ldots, h$; $k = N-h, \ldots, N-1$) by a single point in $x = 0$, endowed with multiplicity $2h+1$. The function obtained in this way is $\overset{\times}{g}_N(x) = (2h+1)u(x) + \sum_{k=h+1}^{N-h-1} u(x - \frac{2\pi}{N}k)$. For any x satisfying $2 \cdot \frac{2\pi h}{N} \leq x \leq 2\pi \cdot (1 - \frac{2h}{N})$ we get

$$\begin{aligned}
\overset{\times}{g}_N(x) - g_N(x) &= \sum_{k=1}^{h} (2u(x) - u(x - \frac{2\pi}{N}k) - u(x + \frac{2\pi}{N}k)) \geq \\
&= \sum_{k=1}^{h} (\frac{2\pi}{N}k)^2 \min(\mid u''(x - \frac{2\pi}{N}k) \mid, \mid u''(x + \frac{2\pi}{N}k) \mid) \geq \\
&\geq \frac{2\pi^2}{N^2} \sum_{k=1}^{h} k^2 \sin \frac{\pi h}{N} \geq \frac{2\pi^2}{N^2} \cdot \frac{\pi h}{N} \cdot \frac{2}{\pi} \cdot \frac{h^3}{3} \geq \frac{4\pi^2}{3N^3} h^4.
\end{aligned}$$

We have $\overset{\times}{g}_N(x) \geq \frac{4}{\pi}N$ in the interval $[2 \cdot \frac{2\pi h}{N}, 2\pi(1 - \frac{2h}{N})]$ if the inequality $\frac{4\pi^2}{3N^3}h^4 > \frac{\pi}{2N}$ holds. This is true for $h = [\frac{1}{2}\sqrt{N}] + 1$ ($N > 4$), yielding $m(S(\omega_N)) \leq 4 \cdot \frac{2\pi}{N}([\frac{1}{2}\sqrt{N}] + 1) < \frac{8\pi}{\sqrt{N}} < \frac{30}{\sqrt{N}}$ for all $N > 4$. For $N \leq 4$ this estimate holds trivially. This proves the assertion. □

One could of course also investigate the set of points z, $\mid z \mid = 1$, for which the reverse inequalities $\prod_{j=1}^{N} \mid z - z_j \mid \geq 1$ and $\sum_{j=1}^{N} \mid z - z_j \mid \geq \frac{4}{\pi}N$ hold. Essentially the same upper bounds can be obtained by choosing an arrangement of points z_j also proposed in [1]. Unfortunately, the energy method does not apply in this case, and we could not get the corresponding lower estimates.

5. Characteristic Function of an Interval

Let $f(x) = \chi_{[0,2\pi\alpha)}(x) - \alpha$, $\alpha \in (0,1)$. If α is rational, $\alpha = \frac{p}{q}$ with p, q coprime, we have $\beta_N(f) \geq \frac{2\pi}{q}$ for all N. On the other hand, it is not difficult to show that $\beta_N(f) = \frac{2\pi}{q}$ holds for N large enough ($N > q^3$ will do). Hence the answer to Problem A is positive if and only if α is irrational. The lower and upper bounds for $\beta_N(f)$ seem to depend both on the number N and on the degree with which α can be approximated by rational numbers. One special case should be noted: if $\frac{p}{q}$ is a rational number satisfying the inequality $0 < \frac{p}{q} - \alpha < \frac{c}{q^\gamma}(c, \gamma > 0)$, then, choosing $\omega_q = \frac{0}{q}, \frac{1}{q}, \ldots, \frac{q-1}{q}$, we see that $\beta_q(f) < \frac{2\pi c}{q^{\gamma-1}}$ holds. If η denotes the type of the irrational number α (as defined f.e. in [3], p.121), we deduce that either $\beta_N(f) < 2\pi N^{-\eta+1+\varepsilon}$ or $\beta_N(-f) < 2\pi N^{-\eta+1+\varepsilon}$ holds for infinitely many N and arbitrary $\varepsilon > 0$. Upper bounds for any N in terms of the type η could be derived from 2(c). We shall not do so, however. One has the intuitive feeling that "small" values of $\beta_n(f)$ entail "large" values for $\beta_n(-f)$. This is true to a certain extent, as the following theorem shows.

Theorem 2. *For any function $f(x) \in L^1_{\text{per}}[0, 2\pi)$, $\int_0^{2\pi} f(x)dx = 0$, and arbitrary natural numbers N, M we have*

$$M \cdot \beta_N(f) + N \cdot \beta_M(-f) \geq 2\pi.$$

Proof. Let $\varepsilon > 0$. Let $\{x_1, \ldots, x_N\}$ and $\{y_1, \ldots, y_M\}$ be tuples of points in $[0, 2\pi)$ such that the relations $m(\{x \in [0, 2\pi) : \sum f(x - x_j) \geq 0\}) \leq (1+\varepsilon)\beta_N(f)$ and $m(\{x \in [0, 2\pi) : \sum f(x - y_k) \leq 0\}) \leq (1 + \varepsilon)\beta_M(-f)$ hold, respectively. Consider the function $\sum_j \sum_k f(x - x_j - y_k)$. We have

$$m(\{x : \sum_{j=1}^{N} \sum_{k=1}^{M} f(x - x_j - y_k) < 0\}) \leq N \cdot m(\{x : \sum_{k=1}^{M} f(x - y_k) < 0\}),$$

hence

$$m(\{x : \sum_{j=1}^{N} \sum_{k=1}^{M} f(x - x_j - y_k) \geq 0\}) \geq 2\pi - N \cdot m(\{x : \sum_{k+1}^{M} f(x - y_k) < 0\}) \geq$$

$$\geq 2\pi - N \cdot m(\{x : \sum_{k=1}^{M} f(x - y_k) \leq 0\}) \geq 2\pi - N(1 + \varepsilon)\beta_M(-f).$$

On the other hand, we have the following inequality:

$$m(x : \sum_{k=1}^{M} \sum_{j=1}^{N} f(x - x_j - y_k) \geq 0) \leq$$

$$\leq M \cdot m(x : \sum_{j=1}^{N} f(x - x_j) \geq 0) \leq M \cdot (1 + \varepsilon)\beta_N(f).$$

Since $\varepsilon > 0$ is arbitrary, the result follows. □

We apply Theorem 2 to the example $f(x) = \chi_{[0,2\pi\alpha)}(x) - \alpha$ with $\alpha = \frac{1}{2}(\sqrt{5} - 1)$. Let $F_0 = 1$, $F_1 = 1$, $F_2 = 2$, ... be the sequence of Fibonacci numbers. For $q = F_{2s}$ we have $-\frac{1}{\sqrt{5}} \cdot \frac{1}{F_{2s}^2} < \frac{F_{2s-1}}{2_{2s}} - \alpha < 0$, hence $\beta_{F_{2s}}(-f) <$

$\frac{2\pi}{\sqrt{5}F_{2s}}$.

By Theorem 2 we get for $F_{2s-2} \leq N \leq F_{2s}$:

$$\beta_N(f) \geq \frac{2\pi}{F_{2s}}(1 - N \cdot \frac{1}{\sqrt{5}F_{2s}}) \geq \frac{2\pi}{F_{2s}}(1 - \frac{1}{\sqrt{5}}) \geq \frac{2\pi}{N} \cdot \frac{F_{2s-2}}{F_{2s}}(1 - \frac{1}{\sqrt{5}}) > \frac{1}{N}.$$

Hence we have our final result: $\beta_N(f) < \frac{2\pi}{\sqrt{5N}}$ for infinitely many values of N and $\beta_N(f) > \frac{1}{N}$ for all N. For the great majority of numbers N, however, the actual value of $\beta_N(f)$ should be much larger than $\frac{1}{N}$.

References

[1] Erdős, Herzog and Piranian, Metric properties of polynomials. *J. Analyse Math.* **6** 125-148 (1958).

[2] Fejes Tóth, On the sum of distances determined by a point set. *Acta Math. Acad. Sci. Hungarica* **7** (1956), 397-401.

[3] Katznelson, *Introduction to Harmonic Analysis.* Wiley and Sons (1968)

[4] Kuipers and Niederreiter, *Uniform distribution of sequences.* Wiley and Sons (1974)

[5] Wagner, On a problem of Erdős, Herzog and Piranian. *to appear in: Acta Math. Hung.*

G. Wagner
Mathematisches Institut A
Pfaffenwaldring 57
D-7000 Stuttgart 80
Federal Republic of Germany

15. Problems

R. Ahlswede: A Problem on Equidistribution

Existence proofs by random selection are very popular in Combinatorics, Information Theory, Complexity Theory etc. We wonder whether they can be replaced by deterministic procedures, which have certain equidistribution properties. Our ideas are not yet precise. We came across the following number theoretical problem, which does not seem to fit into the classical theory of equidistribution.

Let a and b be positive integers such that $a > b$ and a, b are relatively prime. Consider the sets $A_n = \{\sum_{i=1}^{n} \varepsilon_i a^i : \varepsilon_i \in \{0, 1, \ldots, b-1\}\}$. Do the sets $A_n(m) = \{k \in A_n : k \equiv m \mod b^n\}$ satisfy for all $0 \leq m \leq b^n - 1$ $|A_n(m)| b^{-n} = 0(1)$ (or at least $|A_n(m)| b^{-n} = 2^{0(n)}$)?

R. Ahlswede: The Partial Transversal Conjecture

In [1] we studied several source coding problems involving decompositions of $n \times n$-arrays into as few as possible partial transversals such that each transversal has distinct symbols as entries. It is therefore of interest to know the possible lengths of such transversals. In particular we have the following:

Conjecture. Suppose that in an $n \times n$-array no symbol occurs more than n times as an entry, then there exists a partial transversal of length $n-1$ with distinct symbols. The example $\begin{pmatrix} ab \\ ba \end{pmatrix}$ shows that one cannot always expect a transversal of length n.

P. Erdős - J. Pach - J. Spencer

Let G have n vertices and $\frac{1}{2}\binom{n}{2}$ edges. Does there exist a partition of the vertices into two sets A, B of cardinality $n/2$ such that, x be the number of edge of $G|_{A\times B}$,

$$\left|x - \frac{1}{4}\binom{n}{2}\right| > cn^{3/2}$$

Remark. We may assume G has e edges where $\varepsilon\binom{n}{2} < e < (1-\varepsilon)\binom{n}{2}$.

The conjecture is that

$$\left|x - \frac{e}{2}\right| > cn^{3/n}$$

Remark. It is known that there is a set A of cardinality $n/2$ such that, letting y be the number of edges of G/A

$$\left|y - \frac{1}{8}\binom{n}{2}\right| > cn^{3/2}$$

P. Erdős - J. Nešetřil

Let e and f be distinct edges in a graph $G = (V, E)$. Say that e, f are *strongly independent* if they induce a copy of $2k_2$.

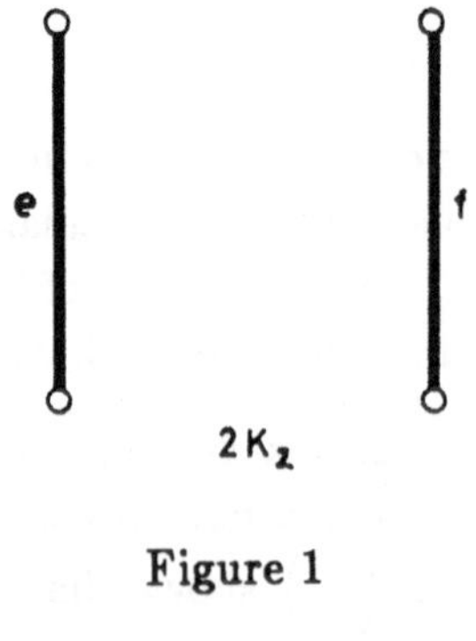

Figure 1

Then define the *strong chromatic index* $X^*(G)$ of G as the least t for which there exists a partition $E = E_1 \cup E_2 \cup \ldots \cup E_t$ so that for each $i = 1, 2, \ldots, t$, any two edges belonging to E_i are strongly independent. This definition leads to the following induced version of Vizing's theorem.

Problem. Find the least $f(k) = t$ so that if $\Delta(G) = k$, then $X^*(G) \leq f(k)$. It is trivial to see that $f(k) \leq 2k^2 - 2k + 1$. On the other hand $f(k) \geq 5k^2/4$ as the following graph shows.

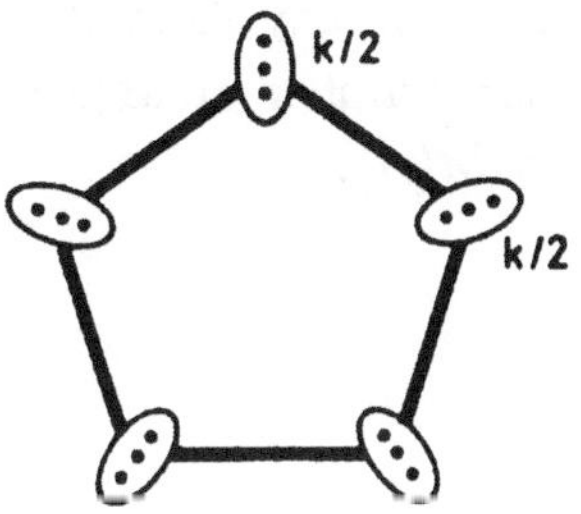

Figure 2

There is one piece of evidence in support of the conjecture $f(k) = 5k^2/4$. Chung, Gyárfás, Troth, and Tuza showed that if $\Delta(G) \leq k$ and G has no induced $2k_2$, then G has at most $5k^2/4$ edges.

L. Babai

Let $\varepsilon > 0$ and $C > 0$. Let G be a finite group and $S \subseteq G, \quad |S| \geq \varepsilon|G|$.

Conjecture. If $|G|$ is large enough then the equation $xy = z$ has a solution in S, provided G satisfies the following two conditions:

(i) G is its own commutator subgroup;

(ii) G has no subgroup of index $\leq C$.

In particular, does the conjecture hold if G is a large *alternating group* (group of even permutations) or a *special linear group* $SL(n, q)$.

R. Fraissé (Communicated by Pierre Duchet)

This (unpublished) problem deals with matrices whose entries are 0 or 1 and having a finite number of rows and columns. A matrix $A = (a_{i,j})$ is *extracted* from another matrix $B = (b_{k,\ell})$ when there exist injective mappings φ and ψ such that $(\forall_{i,j}) \quad a_{i,j} = b_{\varphi(i),\psi(j)}$. When A results from the deletion of a single row of B, the matrix B is named an *extension* of A. Let $M(A)$ denote the class of all matrices from which a given matrix A cannot be extracted.

Conjecture (Wrong form). For any fixed A, there exists an integer $N(A)$ such that every numbers of $M(A)$ with at least $N(A)$ rows admits an extension in $M(A)$.

Conjecture (Weak form). For any fixed A, there exist an infinite set of integers $S(A)$ such that every number of $M(A)$ whose number of rows is in $S(A)$ admits an extension in $M(A)$.

R.L. Graham

Let $W^*(n)$ denote the size of the smallest set of integers X such that any 2-coloring of X faces a monochromatic n-term arithmetic progression. Does $W^*(n)/W(n) \to 0$ as $n \to \infty$? Here, $W(n)$ denotes the ordinary van der Waerden function, i.e. the length of the shortest *interval* of integers with the above arrowing property. It is known, for example, that $W(4) = 34$ while $W^*(4) \leq 30$.

R.L. Graham - N.J.A. Sloane

How small can a non-vanishing sum of n-th roots of unity be? If $f(n)$ denotes the minimum possible value of $|\sum_{i \in I} p_i| \neq 0$, where $\{p_0, p_1, \dots, p_{n-1}\}$ denotes the set of n-th roots of unity and I ranges over all proper subsets of $\{0, 1, \dots, n-1\}$ then it is known that:

(i) $f(n) < c \cdot 2^{-n/4}$;

(ii) $f(n) \geq e^{-cn \log n}$ for a suitable $c > 0$.

The argument showing (i) is non-constructive. The best *construction* known achieves only $n^{-c \log n}$

Problem (1) Find a construction which achieves an exponentially small value.

(**2**) Decide whether (1) or (2) (or something in between) is the "truth".

Note. For $n \leq 30$, computation shows that $f(n)$ is only achieved by sums in which (after multiplying by a suitable root p_j) p_i occurs if and only if its complex conjugate p_{-i} occurs. Is this always true?

J. Nešetřil

Let us call a graph G *co-critical* if:

(i) the edges of G can be 2-colored so as to form no monochromatic triangle;
(ii) this is not possible if *any* additional edge is added to G.
For example, removing an edge from K_6 forms a co-critical graph G_0.

Question. Are there an infinite number of *minimal* co-critical graphs, i.e., co-critical graphs which loose this property when any vertex is deleted? Is G_0 the only one?

Zs. Tuza

If a graph G does not contain $k+1$ edge-disjoint triangles then there can be found a set of at most $2k$ edges which meets all triangles of G.

Algorithms and Combinatorics

Editors: R. L. Graham, B. Korte, L. Lovász

Combinatorial mathematics has substantially influenced recent trends and developments in the theory of algorithms and its applications. Conversely, research on algorithms and their complexity has established new perspectives in discrete mathematics. This new series is devoted to the mathematics of these rapidly growing fields with special emphasis on their mutual interactions.

The series will cover areas in pure and applied mathematics as well as computer science, including: combinatorial and discrete optimization, polyhedral combinatorics, graph theory and its algorithmic aspects, network flows, matroids and their applications, algorithms in number theory, group theory etc., coding theory, algorithmic complexity of combinatorial problems, and combinatorial methods in computer science and related areas.

The main body of this series will be monographs ranging in level from first-year graduate up to advanced state-of-the-art research. The books will be conventionally type-set and bound in hard covers. In new and rapidly growing areas, collections of carefully edited monographic articles are also appropriate for this series. Occasionally there will also be "lecture-notes-type" volumes within the series, published as *Study and Research Texts* in soft cover and camera-ready form. This will be primarily an outlet for seminar notes, drafts of text-books with essential novelty in their presentation, and preliminary drafts of monographs.

Prospective readers of the series ALGORITHMS AND COMBINATORICS include scientists and graduate students working in discrete mathematics, operations research and computer science and their applications.

Volume 1

K. H. Borgwardt

The Simplex Method

A Probabilistic Analysis

1987. 42 figures in 115 separate illustrations. XI, 268 pages. ISBN 3-540-17096-0

Contents: Introduction. - The Shadow-Vertex Algorithm. - The Average Number of Pivot Steps. - The Polynomiality of the Expected Number of Steps. - Asymptotic Results. - Problems with Nonnegativity Constraints. - Appendix. - References. - Subject Index.

Springer-Verlag Berlin Heidelberg New York London Paris Tokyo Hong Kong

Springer